RELATED MATHEMATICS FOR CARPENTERS

SECOND EDITION

P. REBAND

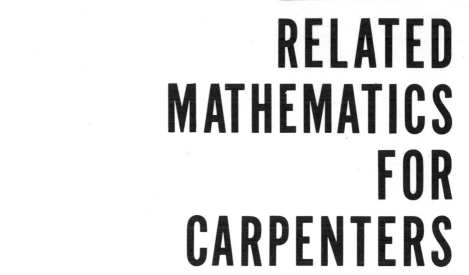

American Technical Society Chicago, 60637

PREFACE TO THE SECOND EDITION

This book provides adequate opportunity for practice on fundamentals and the application of those fundamentals in practical situations. An Answer Key to Self-Check Test questions is also provided at the end of the book so students can check on their individual progress.

The Second Edition of <u>Related</u> <u>Mathematics</u> <u>for</u> <u>Carpenters</u> has been revised to include new sections on accuracy in measuring, the metric system of measurement, missing dimensions and complex areas, and formwork for footings and foundations. New material has also been included on mathematical formulas, areas of triangles and other relevant topics. Several new problems have been added to reflect current practices.

For students who have forgotten the fundamental arithmetic processes, the author has provided page references to the new Third Edition of <u>Practical</u> <u>Mathematics</u>, published by the American Technical Society.

CONTENTS

PART I—BASIC ARITHMETIC

PART III—EVERYDAY PROBLEMS IN THE BUILDING TRADES

From Excavating to Construction
The Ability to Make Simple Calculations
Will Be an Important Factor in Your Success

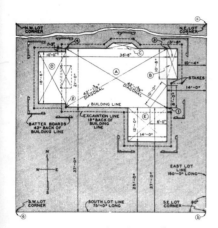

Could you calculate the number of yards of earth to be excavated?

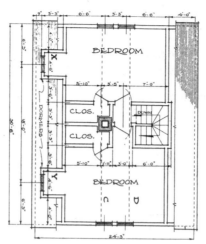

Not all dimensions are shown on a blueprint. Can you find the missing ones by arithmetic?

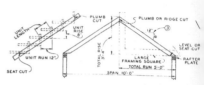

Can you figure the total rise of a roof when you know the pitch and the span

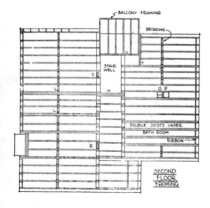

Could you quickly and accurately find the quantity of floor joists that would be needed and their cost?

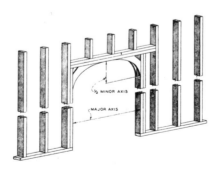

Could you lay out this elliptical arch?

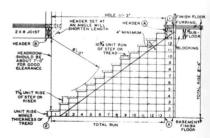

A basic knowledge of mathematics is needed to lay out this stairway. Could you do it?

RELATED MATHEMATICS FOR CARPENTERS

AN IMPORTANT CARPENTERS' TOOL—MATHEMATICS

Many times a day a carpenter must add whole numbers. Adding quantities of material, running feet of lumber, total area of a space, are typical examples. On the right is a floor plan of a house and you want to know the total square feet of area on the first floor. Can you quickly and accurately find the total of 685 sq. ft.? Practice such as this self-check quiz can increase your speed, and improve your accuracy.

SELF-CHECK TEST 1

Addition of Whole Numbers

Study pages 4 to 7 of your textbook, Practical Mathematics, * before preparing this test.

DO NOT SUBMIT THIS SELF-CHECK TEST FOR GRADING. Check your answers with the Answer Key at the end of this workbook.

Add:

(1) 15 in. 18 in. 9 in. 23 in. 54 in. 77 in. *196*	(2) 275 ft. 140 ft. 379 ft. 611 ft. 307 ft. 233 ft. *1945*	(3) 215 sq. in. 1,072 sq. in. 892 sq. in. 711 sq. in. 2,098 sq. in. *4988*	(4) 1,264 cu. in. 789 cu. in. 37 cu. in. 3,701 cu. in. 291 cu. in. *6082*
(5) 1,507 yd. 811 yd. 95 yd. 1,290 yd. *3703*	(6) 27 sq. ft. 1,007 sq. ft. 2,311 sq. ft. 4,284 sq. ft. *7629*	(7) 3,149 sq. yd. 1,003 sq. yd. 693 sq. yd. 2,510 sq. yd. *7355*	(8) $12,500 2,185 4,987 10,090 *$29,762*
(9) 476 lb. 1,117 lb. 805 lb. 2,891 lb. 9,054 lb. *14343*	(10) 3,500 gal. 1,402 gal. 2,917 gal. 2,000 gal. 916 gal. *10,735*	(11) 15,755 cu. yd. 9,689 cu. yd. 12,075 cu. yd. 16,917 cu. yd. 3,645 cu. yd. *58081*	(12) 5,250 B.t.u. 13,725 B.t.u. 8,914 B.t.u. 27,605 B.t.u. 7,692 B.t.u. 35,158 B.t.u. *98344*

*NOTE: PAGE NUMBERS THROUGHOUT THIS BOOK REFER TO THE THIRD EDITION OF PRACTICAL MATHEMATICS, PUBLISHED BY THE AMERICAN TECHNICAL SOCIETY.

Problems	Work Here

(1) What is the total floor space in a basement with a recreation room, store room, and laundry containing 390 sq. ft., 99 sq. ft., and 187 sq. ft. ?

Answer: 676 sq. ft.

187
99
390
_____
676

(2) A carpenter lays 312 sq. ft. of shingles the first day, 328 sq. ft. the second, and 344 sq. ft. the third. How many square feet of shingles did he lay in three days?

Answer: 984 sq. ft.

312
328
344
_____
984

(3) Find the total number of square inches of plywood in four pieces containing 625 sq. in., 1,024 sq. in., 729 sq. in., and 1,225 sq. in.

Answer: 3603 sq. in.

625
1024
729
1225
_____
3603

(4) A contractor paid bills of $2,785, $789, $1,342, $1,009, and $1,809 for materials. What was the total cost of the materials?

Answer: $7734.00

2785
789
1342
1009
1809
_____
7734

(5) In making five excavations the following cubic yards of earth were removed: 5,250; 5,670; 7,565; 11,255; and 35,335. Find the total number of cubic yards of earth removed.

Answer: 65,075

5250
5670
7565
11255
35335
_____
65075

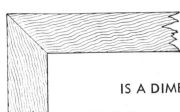

IS A DIMENSION MISSING?—USE SUBTRACTION

If all dimensions were included on a blueprint it would appear very cluttered and hard to read. Many dimensions are found by subtraction as shown below.

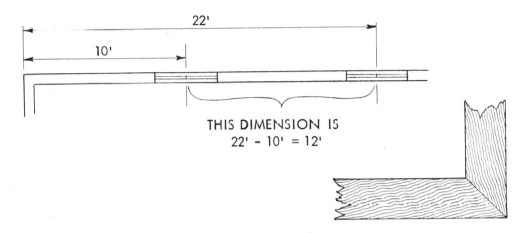

THIS DIMENSION IS
22' - 10' = 12'

SELF-CHECK TEST 2

Subtraction of Whole Numbers

Study pages 7 to 11 of your textbook, Practical Mathematics, before preparing this test.

DO NOT SUBMIT THIS SELF-CHECK TEST FOR GRADING. Check your answers with the Answer Key at the end of this workbook.

Subtract:

(1)   121 in.
       47 in.
      74 in

(2) 1,587 ft.
      609 ft.
     978 ft

(3)   900 yd.
      673 yd.
     227 yd

(4) 1,800 sq. in.
      709 sq. in.
     1,091 sq in

(5) 3,578 sq. ft.
     987 sq. ft.
    2,591 sq ft

(6) 3,092 sq. yd.
    1,007 sq. yd.
   2,085 sq yd

(7) 12,875 cu. yd.
     7,038 cu. yd.
    5,837 cu yd

(8) 16,300 lb.
     3,702 lb.
    12,598 lb

(9) 23,125 cu. ft.
    10,075 cu. ft.
   13,050 cu ft

(10) $14,250
      9,698
     $4,552

(11) 55,875 B.t.u.
     34,996 B.t.u.
    20,879 B.t.u.

(12) $25,000
     11,901
    $13,099

Problem	Work Here

(1) A piece of board 39 in. long was cut from a board 72 in. long. Find the length of the remaining piece.

Answer: *33 in long.*

$$\begin{array}{r} 72 \\ 39 \\ \hline 33 \end{array}$$

(2) How many square feet of plywood remain from an original supply of 10,000 sq. ft. after 6,875 sq. ft. have been used?

Answer: *3,125 sq ft.*

$$\begin{array}{r} 10\,000 \\ 6,875 \\ \hline 3,125 \end{array}$$

(3) It was necessary to use 6,847 bricks from a pile containing 8,750. How many bricks were left?

Answer: *1,903*

$$\begin{array}{r} 8,750 \\ 6,847 \\ \hline 1,903 \end{array}$$

(4) A basement floor space contains 12,250 sq. ft. How much space remains to be painted after 6,835 sq. ft. have been covered?

Answer: *5,415 sq ft.*

$$\begin{array}{r} 12,250 \\ 6,835\ sq\ ft. \\ \hline 5,415 \end{array}$$

(5) A contract for excavating called for the removal of 60,000 cu. yd. of earth. How much remained after removing 7,875 cu. yd.?

Answer: *52,125 cu yd*

$$\begin{array}{r} 60,000 \\ 7,875 \\ \hline 52125 \end{array}$$

MULTIPLICATION—A TIME SAVER

If a carpenter averages 96 square feet of sheathing per hour, how much can he install in 8 hrs.? You could put 96 down 8 times and add, but it is much simpler to multiply.

$$96 \times 8 = 768$$

SHEATHING

SELF-CHECK TEST 3

Multiplication of Whole Numbers

Study pages 11 to 15 of your textbook, Practical Mathematics, before preparing this test.

DO NOT SUBMIT THIS SELF-CHECK TEST FOR GRADING. Check your answers with the Answer Key at the end of this workbook.

Multiply:

(1) 12 in.
42

(2) 52 ft.
15

(3) 248 sq. in.
11

(4) 325 sq. ft.
57

(5) 1,089 cu. in.
26

(6) 2,384 sq. yd.
107

(7) 695 cu. yd.
219

(8) 300 hr.
528

(9) $1,250
54

(10) 1,728 cu. in.
106

(11) 1,250 gal.
875

(12) $16,500
75

Problems	Work Here

(1) What is the total length of 21 pieces of lath each 28 in. long?

Answer: 588

$$\begin{array}{r} 28 \\ 21 \\ \hline 28 \\ 560 \\ \hline \end{array}$$

(2) Determine the total wall area of a square room if each wall has an area of 275 sq. ft.

Answer: 1100 sq ft.

$$\begin{array}{r} 33 \\ 275 \\ 4 \\ \hline 1100 \end{array}$$

(3) A man places 65 lineal ft. of joists per hour. How many feet can he place in 8 hr.?

Answer: 520

$$\begin{array}{r} 4 \\ 65 \\ 8 \\ \hline 520 \end{array}$$

(4) One man can lay 41 sq. ft. of roof shingles per hour. How many square feet can 4 men working at the same rate lay in an 8-hour day?

Answer: 1312

$$\begin{array}{r} 41 \\ 32 \\ \hline 82 \\ 1230 \\ \hline 1312 \end{array}$$

(5) The cost of a new garage is $2,275. Find the cost of 17 garages of this kind.

Answer:

38,675

$$\begin{array}{r} 153 \\ 2275 \\ 17 \\ \hline 15925 \\ 22750 \\ \hline \$38,675 \end{array}$$

MATHEMATICS DEVELOPED TO MEET NEEDS OF BUILDERS

$28 \div 6 = ?$

As far back as a thousand years before Christ, the practical needs of carpentry and construction stimulated the development of mathematics. For many centuries division was done by subtraction as shown at the right. Our long division method was not invented until shortly before the discovery of America.

$$
\begin{array}{r}
28 \\
-6 \\
\hline 22 \\
-6 \\
\hline 16 \\
-6 \\
\hline 10 \\
-6 \\
\hline 4
\end{array}
$$

6 goes into 28 four times with a remainder of 4.

SELF-CHECK TEST 4

Division of Whole Numbers

Study pages 16 to 22 of your textbook, Practical Mathematics, before preparing this test.

DO NOT SUBMIT THIS SELF-CHECK TEST FOR GRADING. Check your answers with the Answer Key at the end of this workbook.

Divide:

(1) $8\overline{)864}$

(2) $9\overline{)8199}$

(3) $5\overline{)15355}$

(4) $12\overline{)6912}$

(5) $36\overline{)3456}$

(6) $75\overline{)26325}$

(7) $144\overline{)4608}$

(8) $256\overline{)26112}$

(9) $200\overline{)211738}$

Problems	Work Here

(1) How many rafters 32 in. long can be cut from a piece of lumber 192 in. long?

Answer:

(2) Determine the number of hours required to lay 760 sq. ft. of sub-flooring at the rate of 80 sq. ft. per hr.

Answer:

(3) How long will it take 4 men to install 12,000 sq. ft. of siding at a rate of 24 sq. ft. per hr. for 1 man?

Answer:

(4) How long will it take to install 3,125 sq. ft. of batt-type insulation at the rate of 125 sq. ft. per hr. ?

Answer:

(5) Find the heat loss per hour from a room if the loss during an 8-hour period is 108,600 B.t.u.

Answer:

Since the time you first learned that one half-dollar and two quarters equalled a dollar, you have used fractions. This lesson will give you practice in the practical use of fractions in carpentry.

$$\frac{1}{2} + \frac{1}{4} + \frac{1}{4} = \frac{4}{4} = 1$$

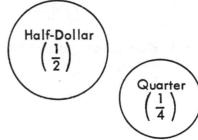

SELF-CHECK TEST 5

Addition of Fractions

Study pages 73 to 77 of your textbook, Practical Mathematics, before preparing this test.

DO NOT SUBMIT THIS SELF-CHECK TEST FOR GRADING. Check your answers with the Answer Key at the end of this workbook.

Problems	Work Here

(1) Add $\frac{1}{4}$ in. and $\frac{1}{2}$ in.

Answer:

(2) Add $\frac{1}{2}$ in. and $\frac{5}{8}$ in.

Answer:

(3) Add $\frac{1}{2}$ in., $\frac{3}{8}$ in., and $\frac{3}{4}$ in.

Answer:

(4) Add $\frac{3}{4}$ yd., $\frac{1}{2}$ yd., and $\frac{2}{3}$ yd.

Answer:

(5) Add $\frac{1}{2}$ sq. ft., $\frac{1}{4}$ sq. ft., $\frac{5}{8}$ sq. ft., and $\frac{2}{3}$ sq. ft.

Answer:

Problems	Work Here

(6) What is the total thickness of three boards $\frac{7}{16}$ in., $\frac{1}{2}$ in., and $\frac{3}{4}$ in. thick?

Answer:

(7) What is the total thickness of a table top made of $\frac{3}{4}$ -inch lumber covered with $\frac{3}{8}$ -inch plywood?

Answer:

(8) A desk top is $\frac{7}{8}$ in. thick. Find the total thickness if it is covered with plate glass $\frac{3}{16}$ in. thick.

Answer:

(9) Find the thickness of a drawing board built of three plies $\frac{1}{2}$ in., $\frac{1}{4}$ in., and $\frac{11}{16}$ in. thick.

Answer:

(10) How thick is a panel built up of four plies $\frac{1}{4}$ in., $\frac{3}{8}$ in., $\frac{5}{16}$ in., and $\frac{7}{32}$ in. thick?

Answer:

Subtraction of Fractions

Study pages 80 to 83 of your textbook, Practical Mathematics, before preparing this test.

DO NOT SUBMIT THIS SELF-CHECK TEST FOR GRADING. Check your answers with the Answer Key at the end of this workbook.

Problems	Work Here

(1) Subtract $\frac{1}{4}$ in. from $\frac{1}{2}$ in.

Answer:

(2) Subtract $\frac{3}{8}$ in. from $\frac{1}{2}$ in.

Answer:

(3) Subtract $\frac{5}{8}$ in. from $\frac{3}{4}$ in.

Answer:

(4) Subtract $\frac{3}{4}$ sq. ft. from $\frac{7}{8}$ sq. ft.

Answer:

(5) Subtract $\frac{2}{3}$ sq. yd. from $\frac{8}{9}$ sq. yd.

Answer:

Problems	Work Here

(6) Find the thickness of a rough board $\frac{1}{2}$ in. thick after $\frac{1}{8}$ in. is taken off by planing on one side.

Answer:

(7) A rough board $\frac{3}{4}$ in. thick has $\frac{1}{16}$ in. taken off by planing on one side. What is its thickness?

Answer:

(8) The total thickness of a table top after being covered with $\frac{1}{4}$-inch linoleum is $\frac{15}{16}$ in. What is the thickness of the top?

Answer:

(9) How much must a $\frac{3}{4}$-inch board be planed to make it the required thickness of $\frac{23}{32}$ in.?

Answer:

(10) The thickness of a piece of three-ply wood is $\frac{11}{16}$ in. It was made by gluing $\frac{1}{8}$-inch pieces on both faces. Find the thickness of the original piece.

Answer:

In many detail drawings whole numbers and fractions must be added to obtain an overall dimension. Could you find the height of the Utility Cabinet shown in section on the right?

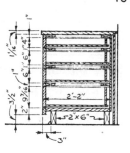

$$3\frac{1}{2}" + 2" + 9\frac{7}{16}" + 1" + 6" + 1" + 6" + 1" + 4" + 1" + 1\frac{1}{16}" = 36"$$

SELF-CHECK TEST 7

Addition of Fractions and Mixed Numbers

Study pages 77 to 80 of your textbook, Practical Mathematics, before preparing this test.

DO NOT SUBMIT THIS SELF-CHECK TEST FOR GRADING. Check your answers with the Answer Key at the end of this workbook.

Problems	Work Here

(1) Add $\frac{3}{2}$ in., $\frac{7}{2}$ in., and $\frac{9}{2}$ in.

Answer:

(2) Add $\frac{1}{4}$ ft., $\frac{9}{4}$ ft., and $\frac{11}{4}$ ft.

Answer:

(3) Add 5 sq. in., $6\frac{1}{2}$ sq. in., and $17\frac{1}{2}$ sq. in.

Answer:

(4) Add $3\frac{1}{4}$ yd., $9\frac{3}{4}$ yd., and 21 yd.

Answer:

(5) Add $\frac{1}{3}$ hr., $3\frac{3}{4}$ hr., and $5\frac{2}{3}$ hr.

Answer:

Problems	Work Here

(6) A platform is built of lumber $2\frac{1}{2}$ in. thick. Find the total thickness if it is covered with boards $\frac{5}{8}$ in. thick.

Answer:

(7) A block is made up by fastening together three pieces $1\frac{3}{16}$ in., $2\frac{3}{8}$ in:, and $3\frac{3}{8}$ in. thick. What is the thickness of the block?

Answer:

(8) What will be the total thickness of an outside wall allowing for $\frac{3}{4}$-inch sheathing, a $3\frac{1}{2}$-inch stud and $\frac{7}{8}$-inch rocklath and plaster?

Answer:

(9) A door is made of 2 pieces of $\frac{5}{8}$-inch and 2 pieces of $\frac{3}{16}$-inch plywood. Find the total thickness of the door.

Answer:

(10) Find the total width required for one floor tile $2\frac{3}{16}$ in. by $2\frac{3}{16}$ in. if $\frac{1}{8}$ in. is allowed on each side for cement.

Answer:

Subtraction of Fractions and Mixed Numbers

Study pages 83 to 89 of your textbook, <u>Practical</u> <u>Mathematics</u>, before preparing this test.

DO NOT SUBMIT THIS SELF–CHECK TEST FOR GRADING. Check your answers with the Answer Key at the end of this workbook.

Problems	Work Here

(1) Subtract $\frac{1}{2}$ in. from $\frac{3}{4}$ in.

    Answer:

(2) Subtract $\frac{3}{8}$ in. from $\frac{13}{16}$ in.

    Answer:

(3) Subtract $\frac{7}{16}$ in. from $\frac{27}{32}$ in.

    Answer:

(4) Subtract $\frac{3}{4}$ in. from $\frac{27}{32}$ in.

    Answer:

(5) Subtract $\frac{7}{8}$ in. from $\frac{29}{32}$ in.

    Answer:

Problems	Work Here

(6) How much longer is a 12d nail than an 8d nail if the lengths are $3\frac{1}{4}$ in. and $2\frac{1}{2}$ in. respectively?

Answer:

(7) Find the difference in width of two pieces of hardwood flooring $\frac{25}{32}$ in. thick by $2\frac{1}{4}$ in. and $\frac{25}{32}$ in. thick by $1\frac{1}{2}$ in.

Answer:

(8) Two pieces of lumber 1 in. by 14 in. and 1 in. by 16 in. each 10 ft. long contain $11\frac{2}{3}$ and $13\frac{1}{3}$ bd. ft. Find the difference in board feet content.

Answer:

(9) By how much does the length of a $\frac{1}{2}$ in. by $1\frac{1}{16}$ in. oblong tile exceed its width?

Answer:

(10) What is the final thickness of a 2-inch thick piece of material if $\frac{3}{16}$ in. is planed off both surfaces?

Answer:

BUILT-UP GIRDERS

Large, solid girders frequently warp or may contain decayed wood, so built-up girders are used. Can you figure the actual thickness of a girder built-up out of three 2 x 8's?

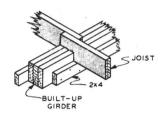

$$1\frac{1}{2}" \times 3 = 4\frac{1}{2}"$$

SELF-CHECK TEST 9

Multiplication of Fractions

Study pages 95 to 109 of your textbook, <u>Practical Mathematics</u>, before preparing this test.

DO NOT SUBMIT THIS SELF-CHECK TEST FOR GRADING. Check your answers with the Answer Key at the end of this workbook.

Problems	Work Here
(1) Multiply $\frac{3}{4}$ in. by $\frac{1}{2}$. Answer:	
(2) Multiply $\frac{5}{8}$ in. by $\frac{1}{4}$. Answer:	
(3) Multiply $\frac{3}{4}$ in. by $\frac{3}{4}$. Answer:	
(4) What is $\frac{1}{2}$ of $\frac{3}{4}$ ft.? Answer:	
(5) What is $\frac{3}{4}$ of $\frac{2}{3}$ ft.? Answer:	

(6)   Twenty-one windows require 3/4
gallon of paint. Find the amount
for seven windows.

Answer:

(7)   If $\frac{1}{2}$ gallon of paint will cover approximately 350 square feet of surface,
find the quantity necessary for 175
square feet.

Answer:

(8)   Sixty sash require $\frac{3}{4}$ gallon of paint
for one coat, determine the amount
for 40 sash.

Answer:

(9)   Three fourths of a gallon of paint is
required for 375 lineal feet of gutters.
How much is needed for 250 lineal feet?

Answer:

(10) If three-fourths gallon of paint covers
600 square feet of radiators with one
coat, find the number of quarts required
for 200 square feet.

Answer:

Division of Fractions

Study pages 112 to 117 of your textbook, <u>Practical</u> <u>Mathematics</u>, before preparing this test.

DO NOT SUBMIT THIS SELF–CHECK TEST FOR GRADING. Check your answers with the Answer Key at the end of this workbook.

Problems	Work Here

(1) Divide $\frac{1}{2}$ in. by 2.

Answer:

(2) Divide $\frac{3}{4}$ ft. by 3.

Answer:

(3) Divide $\frac{5}{9}$ sq. yd. by $\frac{1}{3}$.

Answer:

(4) Divide $\frac{5}{8}$ in. by $\frac{3}{4}$.

Answer:

(5) Divide $\frac{11}{16}$ in. by $\frac{5}{8}$.

Answer:

Problems	Work Here

(6) How many feet are represented by a line $\frac{7}{8}$ in. long if it is drawn so that $\frac{1}{2}$ in. equals 1 ft.?

Answer:

(7) Find the number of pieces of $\frac{3}{16}$-inch plywood necessary to make a piece $\frac{3}{4}$ in. thick.

Answer:

(8) If $\frac{1}{4}$ in. on a blueprint is equal to 1 foot of actual sized dimension, determine the length represented by $\frac{7}{8}$ in.

Answer:

(9) A blueprint is made to a $\frac{1}{8}'' = 1'0''$ scale. How many feet are represented by $\frac{15}{16}$ in.?

Answer:

(10) Determine the number of feet represented by $\frac{11}{16}$ in. if a blueprint is made to a $\frac{1}{8}'' = 1'0''$ scale.

Answer

LAYING OUT RAFTERS

One of the allowances made at the ridge cut of a common rafter is for the thickness of the ridge itself. The common rafter must be shortened by a distance equal to half the thickness of the ridge.

$$\frac{1}{2} \times 1\frac{1}{2}" = \frac{3}{4}"$$

SELF-CHECK TEST 11

Multiplication of Fractions and Mixed Numbers

Study pages 108 to 111 of your textbook, <u>Practical</u> <u>Mathematics</u>, before preparing this test.

DO NOT SUBMIT THIS SELF-CHECK TEST FOR GRADING. Check your answers with the Answer Key at the end of this workbook.

Problems	Work Here

(1) Multiply $\frac{3}{8}$ in. by $\frac{1}{2}$.

Answer:

(2) Multiply $\frac{3}{4}$ ft. by $\frac{4}{9}$.

Answer:

(3) Multiply $2\frac{1}{4}$ in. by 4.

Answer:

(4) Multiply $5\frac{1}{2}$ yd. by $2\frac{3}{4}$.

Answer:

(5) Multiply $10\frac{2}{3}$ ft. by $3\frac{1}{3}$.

Answer:

Problems	Work Here

(6) Find the time required to install a $13\frac{1}{2}$-foot clothes chute if it takes $\frac{1}{2}$ hr. per ft.

Answer:

(7) A board $5\frac{1}{2}$ in. wide is cut to three-fourths of its original width. Find the new width.

Answer:

(8) If $\frac{1}{2}$ hr. of labor is required to place 10 lineal ft. of sills, find the time necessary to place 125 lineal feet.

Answer:

(9) A beam $\frac{5}{8}$ the length of one $24\frac{2}{3}$ ft. long is to be cut. What is its length?

Answer:

(10) A floor board $12\frac{3}{4}$ ft. long is to be cut into two parts one which is two-thirds of this length. How long is this part?

Answer:

Division of Fractions and Mixed Numbers

Study pages 118 to 125 of your textbook, <u>Practical</u> <u>Mathematics</u>, before preparing this test.

DO NOT SUBMIT THIS SELF-CHECK TEST FOR GRADING. Check your answers with the Answer Key at the end of this workbook.

Problems	Work Here

(1) Divide $\frac{1}{2}$ in. by 2.

Answer:

(2) Divide $\frac{3}{4}$ in. by $\frac{1}{2}$.

Answer:

(3) Divide $3\frac{3}{4}$ ft. by 3.

Answer:

(4) Divide $5\frac{5}{8}$ in. by $4\frac{1}{2}$.

Answer:

(5) Divide $22\frac{1}{2}$ sq. yd. by $\frac{3}{5}$.

Answer:

Problems	Work Here

(6) How many shelf boards $3\frac{1}{2}$ ft. long can be cut from a 14-foot board?

Answer:

(7) How many pieces of floor covering $3\frac{1}{4}$ in. wide are needed to cover 39 in. of floor width?

Answer:

(8) How many boards $5\frac{1}{4}$ in. wide will it take to cover a floor $10\frac{1}{2}$ ft. wide?

Answer:

(9) How many supporting columns 7 ft. 6 in. long can be cut from 8 pieces, each $22\frac{1}{2}$ ft. long?

Answer:

(10) How many pieces of $\frac{3}{16}$-inch plywood are there in a stack 4 ft. 3 in. high?

Answer:

RELATED MATHEMATICS FOR CARPENTERS

Examination No. 1

Based on pages 1 to 24 of this workbook

Student's Name_____Student Number_____

Street_____City_____State_____Zip Code_____

Add:	Subtract:	Multiply:	Divide:
(1)　　344 sq. ft. 　1,193 sq. ft. 　　75 sq. ft. 　706 sq. ft. 　521 sq. ft.	(2) 1,500 sq. in. 　　893 sq. in.	(3)　248 sq. in. 　207	(4) 144)‾6048

Problems	Show Work Here

(5) Four pieces of plywood contain 1,089 sq. in., 374 sq. in., 1,375 sq. in., and 2,709 sq. in. Find the total number of square inches of plywood in the four pieces.

Answer:

(6) A certain job required the application of 2,300 sq. ft. of sheathing. How many square feet remained to be done after 1,288 sq. ft. had been applied?

Answer:

CUT OFF HERE

Problems	Show Work Here

(7) A job requires 58 pieces of 2 by 4's each 17 ft. long. What is the total length?

Answer:

(8) How long will it take two men to apply 12,288 sq. ft. of siding at the rate of 96 sq. ft. per hr. for 4 men?

Answer:

(9) Add $\frac{3}{4}$ in., $\frac{7}{8}$ in., and $\frac{9}{16}$ in.

Answer:

(10) Subtract $\frac{39}{64}$ in. from $\frac{7}{8}$ in.

Answer:

(11) Multiply $\frac{15}{16}$ in. by $\frac{2}{3}$.

Answer:

(12) Divide $\frac{13}{16}$ in. by 2.

Answer:

Student's Name_____Student Number_____

Problems	Show Work Here

(13) Find the total thickness of three boards: $\frac{5}{8}$ in., $\frac{11}{16}$ in., and $\frac{3}{4}$ in. thick.

Answer:

(14) Which is thicker, 12 pieces of $\frac{1}{4}$-inch plywood or 5 pieces of $\frac{11}{16}$-inch plywood?

Answer:

(15) How much must a board which is $\frac{7}{8}$ in. thick be planed to make it $\frac{25}{32}$ in. thick?

Answer:

(16) How many pieces of $\frac{13}{16}$-inch lumber are there in a stack 52 in. high?

Answer:

Problems	Show Work Here

(17) Find the total thickness of a block of wood made by fastening together three pieces: $2\frac{7}{8}$ in., $1\frac{11}{16}$ in., and $3\frac{1}{4}$ in. thick.

Answer:

(18) What is the total thickness in feet of 24 boards each $1\frac{7}{8}$ in. thick?

Answer:

(19) Shingles are applied to the side of a building so that $7\frac{1}{2}$ in. are exposed. What will be the height in feet of the surface covered if there are 22 courses?

Answer:

(20) Find the number of boards $3\frac{7}{8}$ in. wide required to cover a space 5 ft. 2 in. wide.

Answer:

Addition of Decimals

Study pages 137 to 139 of your textbook, <u>Practical</u> <u>Mathematics</u>, before preparing this test.

DO NOT SUBMIT THIS SELF–CHECK TEST FOR GRADING. Check your answers with the Answer Key at the end of this workbook.

Problems	Work Here

(1) Add .375 in. and .25 in.

    Answer:

(2) Add .875 in. and .625 in.

    Answer:

(3) Add 4.125 in., 3.625 in.,
    6.25 in., and 2.5 in.

    Answer:

(4) Add .0625 in., .3125 in.,
    .5625 in., and .6875 in.

    Answer:

(5) Add $33.18, $25.96, $13.27,
    $47.76, and $54.06.

    Answer:

Problems	Work Here

(6) A table top 1.875 in. thick is
covered with linoleum .187 in.
thick. Find the total thickness
of the top.

Answer:

(7) A piece of three-plywood is made
up of an inner piece .375 in. thick
covered by two pieces each .250
in. thick. What is its thickness?

Answer:

(8) Find the outside diameter of a
wrought-iron pipe if the thickness
of the metal is .280 in. and the
inside diameter is 6.065 in.

Answer:

(9) A contractor paid bills of $132.16,
$79.34, $1,004.07, $507.83, and
$1,213.56 for materials. Find the
total amount of the bills.

Answer:

(10) After completing six bungalows, the
contractor found the costs to be
$13,125.64, $14,073.29, $14,978.87,
$13,409.11, $15,084.16, and $15,490.37.
Find the total cost.

Answer:

Subtraction of Decimals

Study pages 137 to 139 of your textbook, <u>Practical</u> <u>Mathematics</u>, before preparing this test.

DO NOT SUBMIT THIS SELF–CHECK TEST FOR GRADING. Check your answers with
the Answer Key at the end of this workbook.

Problems	Work Here

(1) Subtract 1.25 in. from 2.5 in.

    Answer:

(2) Subtract .375 in. from .625 in.

    Answer:

(3) Subtract 1.875 in. from 3 in.

    Answer:

(4) Subtract 143.4 ft. from 156.3 ft.

    Answer:

(5) Subtract $68.73 from $94.11.

    Answer:

Problems	Work Here

(6) How much greater is the diameter of a 10d nail than a 6d nail if the diameters are .148 in. and .113 in. respectively?

Answer:

(7) By how much does the length of a lot exceed its width if the length is 195.2 ft. and the width 35.8 ft.?

Answer:

(8) Find the difference in weight per square foot for two pieces of tempered hardboard if the weights are .79 and 1.10 lb. per sq. ft.

Answer:

(9) A contract was accepted for $25,000. The total cost of material and labor was $22,708.92. How much was the profit on the contract?

Answer:

(10) A contractor paid bills of $131.25, $278.16, and $78.29 for materials and labor on a certain job. He accepted the contract for $550. What was his profit?

Answer:

Multiplication of Decimals

Study pages 140 to 144 of your textbook, <u>Practical</u> <u>Mathematics</u>, before preparing this test.

DO NOT SUBMIT THIS SELF-CHECK TEST FOR GRADING. Check your answers with the Answer Key at the end of this workbook.

Problems	Work Here

(1) Multiply 1.25 in. by 3.

   Answer:

(2) Multiply .625 in. by 5.

   Answer:

(3) Multiply 2.375 in. by 2.5.

   Answer:

(4) Multiply 3.75 in. by 5.25.

   Answer:

(5) Multiply $87.96 by 23.5.

   Answer:

Problems	Work Here

(6) Find the weight of 500 sq. ft. of $\frac{1}{8}$-inch asphalt tile if the average weight per square foot is 1.16 lb.

Answer:

(7) What is the weight of 725 sq. ft. of $\frac{3}{16}$-inch rubber tile if the average weight per square foot is 1.86 lb.?

Answer:

(8) The weight of $\frac{3}{8}$-inch plywood is 1.10 lb. per sq. ft. What is the weight of 1,152 sq. ft.?

Answer:

(9) Determine the weight of a 3 ft. by 6 ft. pane of $\frac{1}{4}$-inch plate glass if the weight per square foot is 3.27 lb.

Answer:

(10) If one square foot of $4\frac{1}{2}$-inch wall requires 6.31 firebrick, find the approximate number for 24 sq. ft.

Answer:

Division of Decimals

Study pages 144 to 149 of your textbook, <u>Practical</u> <u>Mathematics</u>, before preparing this test.

DO NOT SUBMIT THIS SELF–CHECK TEST FOR GRADING. Check your answers with the Answer Key at the end of this workbook.

Problems	Work Here

(1) Divide 7.5 in. by 5.

Answer:

(2) Divide 15 in. by .75.

Answer:

(3) Divide 8.75 in. by 1.25.

Answer:

(4) Divide 5.625 in. by .0625.

Answer:

(5) Divide $95.36 by 3.2.

Answer:

Problems	Work Here

(6) How many pieces of plywood each
.375 in. thick are there in a stack
20.25 in. high?

Answer:

(7) A stack of table tops each 1.875 in.
thick is 3 ft. 9 in. high. How many
table tops are there in the stack?

Answer:

(8) A workman received a pay check
for $250. How many hours did he
work if the wage rate was $6.25
per hr. ?

Answer:

(9) A check for $433.80 was received
by a workman. Find the number
of hours which he worked if the
rate was $6.025 per hr.

Answer:

(10) A check for $356.50 was received
for 62 hr. of work. Find the hourly
rate.

Answer:

Changing Common Fractions to Decimals

Study page 148 of your textbook, <u>Practical</u> <u>Mathematics</u>, before preparing this test.

DO NOT SUBMIT THIS SELF–CHECK TEST FOR GRADING. Check your answers with
the Answer Key at the end of this workbook.

Problems	Work Here

(1) Change $\frac{3}{4}$ to a decimal.

     Answer:

(2) Change $\frac{5}{8}$ to a decimal.

     Answer:

(3) Change $\frac{3}{16}$ to a decimal.

     Answer:

(4) Change $\frac{25}{32}$ to a decimal.

     Answer:

(5) Change $\frac{45}{64}$ to a decimal.

     Answer:

Problems	Work Here

(6) About $2\frac{3}{4}$ lb. of nails are required for each 100 square feet of 1 x 8'' subflooring nailed to joists 16'' from center to center. Write this weight in decimal form.

Answer:

(7) The finished width in inches of some shiplap boards is $1\frac{3}{8}$ in. Write the width in decimal form.

Answer:

(8) The approximate thickness of a piece of asphalt tile is $\frac{3}{16}$ in. Write this thickness in decimal form.

Answer:

(9) The nominal thickness of a piece of oak flooring is $\frac{13}{16}$ in. Find this thickness in decimal form.

Answer:

(10) The actual thickness of a piece of oak flooring is $\frac{11}{32}$ in. What is the thickness written as a decimal?

Answer:

Changing Decimals to Common Fractions

Study the illustrative example on page 150 of your textbook, <u>Practical Mathematics</u>, before preparing this test.

DO NOT SUBMIT THIS SELF-CHECK TEST FOR GRADING. Check your answers with the Answer Key at the end of this workbook.

Problems	Work Here

(1) Change .28 to the nearest 4th.

Answer:

(2) Change .869 to the nearest 8th.

Answer:

(3) Change .6457 to the nearest 8th.

Answer:

(4) Change .44325 to the nearest 16th.

Answer:

(5) Change .616832 to the nearest 8th.

Answer:

Problems	Work Here

(6) The weight of $\frac{1}{4}$-inch plywood is .70 lb. per sq. ft. Change .70 to the nearest sixteenth.

Answer:

(7) A piece of material is .381 in. thick. Change this thickness to the nearest eighth of an inch.

Answer:

(8) A piece of plywood is .614 in. thick. What is its thickness to the nearest eighth of an inch?

Answer:

(9) Find the approximate thickness in sixteenths of an inch of a piece of linoleum .187 in. thick.

Answer:

(10) A piece of lining felt is .040 in. thick. What is the thickness to the nearest thirty-second of an inch?

Answer:

Significant Number of Figures and Accuracy of Measurement

DO NOT SUBMIT THIS SELF-CHECK TEST FOR GRADING. Check your answers
with the Answer Key at the end of this workbook.

Information

       In counting items an exact number can be determined. Measuring, how-
ever, is an approximation or comparison at best. Modern measuring devices,
capable of measuring remarkably small and accurate amounts, are often un-
necessary or even undesirable for practical carpentry. A cabinetworker
must, of course, be more accurate than a maker of concrete forms. But
neither requires the precision of a machinist.

       The accuracy of a number is shown by its significant number of figures.
A significant figure is one known to be reasonably accurate.
For Example:

481	(three significant figures)
683.7	(four significant figures)
32.23	(four significant figures)
2008.1	(five significant figures)
99.378	(five significant figures)
0.00056	(two significant figures—Zeros used to mark decimal places are not considered significant figures unless preceded by a non-zero digit)
2.00230	(six significant figures)

      In a mathematical operation, the answer cannot be more accurate than
the data used to acquire the answer. For example, a room is measured in
feet and inches and converted to feet and tenths of a foot. It is found to be
32.3 ft. long by 14.7 ft. wide. Multiplying these two numbers to find the
area of the floor gives the answer as 474.81 sq. ft.

           32.3 ft.   (3 significant figures)
        × 14.7 ft.   (3 significant figures)
       474.81 sq. ft. (5 significant figures)

       This number, with its five significant figures, implies an accuracy to
a hundredth of a square foot, but this is clearly impossible because the
measurements used to arrive at this figure were not that accurate themselves,
and no result of calculation can be more accurate than the data it is based on.
The answer should be stated no closer than 474.8 sq. ft. with some doubt as
to the last digit. An answer of 475 sq. ft. is probably close enough, as it has
three significant figures—the same degree of precision as the original data.
       This fallacy of false accuracy is even more pronounced in calculating
volumes because one more multiplication is used. In the above example of
the room, suppose we also measure the height and find it to be 11.4 ft. By
straightforward arithmetic we find the volume to be 5412.834 cu. ft., a
number with seven significant figures and quite remarkable considering that
none of the data approached that degree of accuracy. A better answer would
be 5413 cu. ft. with some doubt as to the last digit. An even more sensible
answer would be 5410 cu. ft.

Sample Problem 1

Round off according to the information given:

27.3375 sq. ft. × 8.34 ft. = 227.994750 cu. ft.

Since three significant figures is the least accurate of measurements, the answer should be rounded to three significant figures.

227.994750 = 228 cu. ft.

Sample Problem 2

Round off according to the information given:

55.000000 ÷ 15.5 = 3.3333

15.5 contains three significant figures, therefore the answer should contain no more than three significant figures.

3.3333 = 3.33

When adding or subtracting, figure the number of places to the right of the decimal point according to the least accurate measurement used.

Sample Problem 3

Add the following and round off the answer.

```
  2065.328
    42.7
+  203.16
  2311.188  =  2311.2
```

Problems	Work Here

Perform the following mathematical operations, rounding off the answer according to the accuracy of information given.

(1) 100.18 × 64.58   Answer:

(2) 15 ft. 6 inches × 4 ft. 3 in.   Answer:

(3) 227.994750 ÷ 8.34   Answer:

(4) 32.6 ft. × 66.84 ft. × 10 ft.   Answer:

(5) 273.2 - 1.03   Answer:

RELATED MATHEMATICS FOR CARPENTERS

Examination No. 2

Based on pages 29 to 42 in this workbook

Student's Name_____ Student Number_____

Street_____ City_____ State_____ Zip Code_____

Add:	Subtract:	Multiply:	Divide:

(1) 6.375 in.
     3.5   in.
     9.625 in.
    14.75 in.

(2) 10.0165 in.
      7.875 in.

(3) 10.125 in.
       2.7

(4) .375$\overline{)54}$

Problems	Show Work Here

(5) A piece of wood 2.1875 in. thick is covered on one side with a piece of plywood .375 in. thick and on the other side with a piece .250 in. thick. Find the total thickness?

Answer:

(6) A piece of lumber 2.75 in. thick is planed until its thickness is 2.3125 in. How much was removed by planing?

Answer:

CUT OFF HERE

Problems	Show Work Here

(7) What is the total thickness of 12
pieces of plywood .250 in. thick
and 21 pieces .375 in. thick?

Answer:

(8) How many pieces of plywood .375
in. thick are there in a stack 27 in.
high?

Answer:

(9) Change $\frac{7}{8}$ to a decimal fraction.

Answer:

(10) Change $\frac{7}{16}$ to a decimal fraction.

Answer:

(11) Find the decimal fraction equal to
$\frac{21}{32}$.

Answer:

Problems	Show Work Here

(12) Change .369 to the nearest 8th.

    Answer:

(13) Change .2432 to the nearest 4th.

    Answer:

(14) Change .59375 to the nearest 32nd.

    Answer:

(15) What is the thickness in common
    fraction form of a piece of wood
    .750 in. thick?

    Answer:

CUT OFF HERE

Problems	Show Work Here

(16) How much thicker are 20 sheets of
plywood .375 in. thick than 25
sheets .250 in. thick?

Answer:

(17) Plate glass $\frac{1}{4}$ in. thick weighs 3.27
lb. per sq. ft. Find the weight of 6
panes $2\frac{1}{2}$ ft. by 5 ft.

Answer:

(18) A workman worked for 42 hr. at the
rate of $4.80 per hr. How much did
he earn?

Answer:

(19) Multiply 8.625 in. by .31831,
rounding off the answer according to the information given.

Answer:

(20) How many square feet of $\frac{3}{8}$-inch plywood are there in a stack weighing
698.5 lb. if it weighs 1.10 lb. per
square foot?

Answer:

Percentage

Study pages 161 to 181 of your textbook, <u>Practical</u> <u>Mathematics</u>, before preparing this test.

DO NOT SUBMIT THIS SELF-CHECK TEST FOR GRADING. Check your answers with the Answer Key at the end of this workbook.

To determine the amount of material for various jobs it is often necessary to add a certain amount to allow for waste in cutting. If an allowance is to be made, it is important to know how to calculate the amount of material required.

Sample Problem 1

Find the amount of roof boards to be ordered for a job if it is estimated that 2,500 board ft. are necessary and 20 per cent is allowed for waste.

Instructions	Computations
(1) Find 20 per cent of 2,500.	(1) 2,500 × .20 = 500.00
(2) Add 500 to 2,500.	(2) 2,500 + 500 = 3,000

The amount to be ordered is 3,000 board ft.

Sample Problem 2

How many square feet of floor felt should be ordered for a job if the area of the floor is 1,450 ft. and 10 per cent is allowed for waste?

Instructions	Computations
(1) Find 10 per cent of 1,450.	(1) 1,450 × .10 = 145.00
(2) Add 145 to 1450.	(2) 1,450 + 145 = 1,595

The amount to be ordered is 1,595 sq. ft.

Problems	Work Here

(1) The total area to be covered with insulation board is 1,200 sq. ft. How much should be ordered if 6 per cent is allowed for waste?

Answer:

Problems	Work Here

(2) It is found that about 56 lb. of material for drip caps is needed for a certain job. How much should be provided if 25 per cent is allowed for waste?

Answer:

(3) In making mortar, lime is used in an amount equal to 10 per cent of the cement. How much lime is necessary if 990 lb. of cement are used?

Answer:

(4) For subflooring diagonally applied, add 30 per cent to the calculated amount. How much must be allowed for 375 sq. ft. of floor space?

Answer:

(5) It is found that 10,000 facebricks are required for a certain job if laid with $\frac{1}{2}$-inch joints. How many more bricks would be required if $\frac{1}{4}$-inch joints, for which 14 per cent must be added, are used?

Answer:

Denominate Numbers

Study pages 207 to 244 of your textbook, <u>Practical</u> <u>Mathematics</u>, before preparing this test.

DO NOT SUBMIT THIS SELF-CHECK TEST FOR GRADING. Check your answers with the Answer Key at the end of this workbook.

Problems	Work Here

(1) Find the number of inches in $2\frac{3}{4}$ ft.

Answer:

(2) How many feet are there in 54 in.?

Answer:

(3) Determine the number of yards in 96 in.

Answer:

(4) How many square feet are there in 1,224 sq. in.?

Answer:

(5) Find the number of square feet in $10\frac{2}{3}$ sq. yd.

Answer:

Problems	Work Here

(6) How many cubic yards are there in 4,320 cu. ft.?

Answer:

(7) Find the number of cubic yards in 2,740.5 cu. ft.

Answer:

(8) One gallon equals 231 cu. in. Find the number of gallons in 3 cu. ft.

Answer:

(9) Find the weight of $12\frac{1}{2}$ cu. ft. of water if one cubic foot weighs 62.5 lb.

Answer:

(10) How many tons are there in 2,450 lb. of crushed rock if one ton weighs 2000 lbs. ?

Answer:

Problems	Work Here

(11) How many inches are there in 1 yd. $2\frac{1}{2}$ ft. 7 in. ?

Answer:

(12) Determine the number of square feet in a plot of land $3\frac{1}{3}$ sq. yd. $25\frac{1}{2}$ sq. ft.

Answer:

(13) Find the total length of three boards whose lengths are 4 ft. 5 in., 6 ft. 2 in., and 5 ft. $7\frac{1}{2}$ in.

Answer:

(14) A piece 7 ft. 11 in. long is cut from a board 13 ft. 2 in. long. Find the length of the remaining piece.

Answer:

(15) How much earth remains to be re-moved from an excavation estimated to contain 800 cu. yd. after 244 cu. yd. 18 cu. ft., and 356 cu. yd. 24 cu. ft. have been removed?

Answer:

Problems	Work Here

(16) Find the total length of six pieces of edging each 3 ft. $3\frac{1}{2}$ in. long.

Answer:

(17) Find the average time for each carpenter if it takes four of them 12 hr. and 36 min. to complete a job.

Answer:

(18) Change 77°F. to C.

$$°C = (°F - 32°) \times \frac{5}{9}.$$

Answer:

(19) Change 115°C. to F.

$$°F = (°C \times \frac{9}{5}) + 32.$$

Answer:

(20) What would a 2 × 4 ($1\frac{1}{2}'' \times 3\frac{1}{2}''$) measure in the metric system?

Answer:

SELF-CHECK TEST 22

Metric System of Measurement

Study pages 207 to 224 and 242 to 243 in your textbook Practical Mathematics, before preparing this test.

DO NOT SUBMIT THIS SELF-CHECK TEST FOR GRADING. Check your answers with the Answer Key at the end of this workbook.

Information

The metric system is based on the number 10. All units of weight, lineal measure and volume are based on the number 10. The basic unit of length is the meter, which is a bit longer than a yard. Rather than dividing by three to obtain the number of feet or by 36 to obtain the number of inches, one divides by the power of ten. For example, a meter has 1000 millimeters, 100 centimeters, or 10 decimeters. Ten meters equals a deckameter, 100 meters is a hectometer and 1000 meters is equivalent to a kilometer.

These prefixes are common in all units of measure in the metric system. Conversions from one unit to another are made by moving the decimal point either to the right or to the left to increase or decrease the amount.

For Example:

10 millimeters (mm) = 1 centimeter (cm)
= 0.1 decimeter (dm)
= 0.01 meter (m)
= 0.001 decameter (Dm)
= 0.0001 hectameter (Hm)
= 0.00001 kilometer (Km)
= 0.000001 myriameter (Mm)

1 myriameter (Mm) = 10 kilometers (Km)
= 100 hectameters (Hm)
= 1000 decameters (Dm)
= 10000 meters (m)
= 100000 decimeters (dm)
= 1000000 centimeters (cm)

If the co-existence of two systems seems inconvenient, as it is, remember that in respect to world wide agreement the United States is the exception. In changing from one system to another there is an awkward period caused by the fact that the units in one are not evenly divisible by those of the other, and this awkwardness continues until one can think entirely in terms of the new system. For example, 300 mm (corresponding to 11.81 in.) is nearly the same as one foot and serves essentially the same purpose. It is an inconvenient number only while it still has to be "translated." When once the metric units become familiar they have their own reality and avoid the complications of calculating with common fractions. For the awkward period of changeover perhaps the easiest and most certain way is to use conversions tables such as the following designed to convert fractional inches, inches, and feet into millimeters.

TABLE 1

Fractional Inch—Millimeter and Foot—Millimeter Conversion Tables
(Based on 1 inch = 25.4 millimeters, exactly)*

FRACTIONAL INCH TO MILLIMETERS

In.	Mm.	In.	Mm.	In.	Mm.	In.	Mm.
1/64	0.397	17/64	6.747	33/64	13.097	49/64	19.447
1/32	0.794	9/32	7.144	17/32	13.494	25/32	19.844
3/64	1.191	19/64	7.541	35/64	13.891	51/64	20.241
1/16	1.588	5/16	7.938	9/16	14.288	13/16	20.638
5/64	1.984	21/64	8.334	37/64	14.684	53/64	21.034
3/32	2.381	11/32	8.731	19/32	15.081	27/32	21.431
7/64	2.778	23/64	9.128	39/64	15.478	55/64	21.828
1/8	3.175	3/8	9.525	5/8	15.875	7/8	22.225
9/64	3.572	25/64	9.922	41/64	16.272	57/64	22.622
5/32	3.969	13/32	10.319	21/32	16.669	29/32	23.019
11/64	4.366	27/64	10.716	43/64	17.066	59/64	23.416
3/16	4.762	7/16	11.112	11/16	17.462	15/16	23.812
13/64	5.159	29/64	11.509	45/64	17.859	61/64	24.209
7/32	5.556	15/32	11.906	23/32	18.256	31/32	24.606
15/64	5.953	31/64	12.303	47/64	18.653	63/64	25.003
1/4	6.350	1/2	12.700	3/4	19.050	1	25.400

INCHES TO MILLIMETERS

In.	Mm.	In.	Mm.	In.	Mm.	In.	Mm.	In.	Mm.	In.	Mm.
1	25.4	3	76.2	5	127.0	7	177.8	9	228.6	11	279.4
2	50.8	4	101.6	6	152.4	8	203.2	10	254.0	12	304.8

FEET TO MILLIMETERS

Ft.	Mm.	Ft.	Mm.	Ft.	Mm.	Ft.	Mm.	Ft.	Mm.
100	30,480	10	3,048	1	304.8	0.1	30.48	0.01	3.048
200	60,960	20	6,096	2	609.6	0.2	60.96	0.02	6.096
300	91,440	30	9,144	3	914.4	0.3	91.44	0.03	9.144
400	121,920	40	12,192	4	1,219.2	0.4	121.92	0.04	12.192
500	152,400	50	15,240	5	1,524.0	0.5	152.40	0.05	15.240
600	182,880	60	18,288	6	1,828.8	0.6	182.88	0.06	18.288
700	213,360	70	21,336	7	2,133.6	0.7	213.36	0.07	21.336
800	243,840	80	24,384	8	2,438.4	0.8	243.84	0.08	24.384
900	274,320	90	27,432	9	2,743.2	0.9	274.32	0.09	27.432
1,000	304,800	100	30,480	10	3,048.0	1.0	304.80	0.10	30.480

* American Standard Practice for Industrial Use (ANSI B48.1)

Example 1: Find millimeter equivalent of 293 feet, 5 47/64 inches.

200 ft	= 60,960.	mm
90 ft	= 27,432.	mm
3 ft	= 914.4	mm
5 in.	= 127.0	mm
47/64 in. =	18.653	mm
293 ft, 5 47/64 in. =	89,452.053	mm

Example 2: Find millimeter equivalent of 71.86 feet.

70. ft =	21,336.	mm
1. ft =	304.8	mm
.80 ft =	243.84	mm
.06 ft =	18.288	mm
71.86 ft =	21,902.928	mm

From Machinery's Handbook, 19th Edition, Industrial Press. New York, N.Y.

Another method, particularly useful where desk calculators or slide rules can be used, is to multiply by constant factors to change from U.S. to metric measurements or divide by the same factors to change from metric to U.S. measurements.

To Convert from Metric to U.S. Units:

1. Multiply by the factor shown in Table 2.
2. Use the resulting quantity "rounded off" to the number of decimal digits needed for practical application.
3. Wherever practical in semi-precision measurements, convert the decimal part of the number to the nearest common fraction.

Sample Problem 1

How many inches are there in 200 millimeters?

TABLE 2. CONVERSION OF METRIC TO ENGLISH UNITS

LENGTHS:		WEIGHTS:	
1 MILLIMETER (MM)	= 0.03937 IN.	1 GRAM (G)	= 0.03527 OZ (AVDP)
1 CENTIMETER (CM)	= 0.3937 IN.	1 KILOGRAM (KG)	= 2.205 LBS
1 METER (M)	= 3.281 FT OR 1.0937 YDS	1 METRIC TON	= 2205 LBS
1 KILOMETER (KM)	= 0.6214 MILES	**LIQUID MEASUREMENTS:**	
AREAS:		1 CU CENTIMETER (CC)	= 0.06102 CU IN.
1 SQ MILLIMETER	= 0.00155 SQ IN.	1 LITER (= 1000 CC)	= 1.057 QUARTS OR 2.113 PINTS OR 61.02 CU INS.
1 SQ CENTIMETER	= 0.155 SQ IN.	**POWER MEASUREMENTS:**	
1 SQ METER	= 10.76 SQ FT OR 1.196 SQ YDS	1 KILOWATT (KW)	= 1.341 HORSEPOWER
VOLUMES:		**TEMPERATURE MEASUREMENTS:**	
1 CU CENTIMETER	= 0.06102 CU IN.	TO CONVERT DEGREES CENTIGRADE TO DEGREES FAHRENHEIT, USE THE FOLLOWING FORMULA: DEG F = (DEG C x 9/5) + 32	
1 CU METER	= 35.31 CU FT OR 1.308 CU YDS		

SOME IMPORTANT FEATURES OF THE CGS SYSTEM ARE:

1 CC OF PURE WATER = 1 GRAM. PURE WATER FREEZES AT 0 DEGREES C AND BOILS AT 100 DEGREES C.

(1) Multiply by the factor in Table 2.

$$200 \times 0.03937 = 7.87400 \text{ inches}$$

(2) Round off for practical application.

$$7.87400 = 7.87 \text{ inches}$$

(3) Convert decimal portion to nearest fraction.

$$7.87 \text{ inches} = 7\frac{7}{8} \text{ inches}$$

To Convert from U.S. to Metric Units:

1. If the U.S. measurement is expressed in fractional form, change this to an equivalent decimal form.

2. Multiply this quantity by the factor shown in Table 3.

3. Round off the result to the precision required.

TABLE 3. CONVERSION OF ENGLISH TO METRIC UNITS

LENGTHS:		WEIGHTS:	
1 INCH	= 2.540 CENTIMETERS	1 OUNCE (AVDP)	= 28.35 GRAMS
1 FOOT	= 30.48 CENTIMETERS	1 POUND	= 453.6 GRAMS OR 0.4536 KILOGRAM
1 YARD	= 91.44 CENTIMETERS OR 0.9144 METERS	1 (SHORT) TON	= 907.2 KILOGRAMS
1 MILE	= 1.609 KILOMETERS	**LIQUID MEASUREMENTS:**	
AREAS:		1 (FLUID) OUNCE	= 0.02957 LITER OR 28.35 GRAMS
1 SQ IN.	= 6.452 SQ CENTIMETERS	1 PINT	= 473.2 CU CENTIMETERS
1 SQ FT	= 929.0 SQ CENTIMETERS OR 0.0929 SQ METER	1 QUART	= 0.9463 LITER
1 SQ YD	= 0.8361 SQ METER	1 (US) GALLON	= 3785 CU CENTIMETERS OR 3.785 LITERS
VOLUMES:		**POWER MEASUREMENTS:**	
1 CU IN.	= 16.39 CU CENTIMETERS	1 HORSEPOWER	= 0.7457 KILOWATT
1 CU FT	= 0.02832 CU METER	**TEMPERATURE MEASUREMENTS:**	
1 CU YD	= 0.7646 CU METER	TO CONVERT DEGREES FAHRENHEIT TO DEGREES CENTIGRADE, USE THE FOLLOWING FORMULA: DEG C = 5/9 (DEG F - 32)	

Relatively small measurements, such as 17.3 cm, are generally expressed in equivalent millimeter form. In this example the measurement would be read as 173 mm.

Sample Problem 2

A lot $1,531\frac{1}{4}$ sq. ft. is equivalent to how many square meters?

(1) Change feet to decimal form.

$1,531\frac{1}{4}$ = 1531.25 square feet

(2) Multiply by the factor in Table 3.

1531.25 × 0.0929 = 142.253125 square meters

(3) Round off the result.

142.253125 meters = 142.25 meters

Temperature Conversion

To change from degrees Centigrade (metric system) to degrees Fahrenheit, multiply the Centigrade degrees by $\frac{9}{5}$ and add 32.

Sample Problem 3

Change 75° Centigrade to Fahrenheit temperature.

$(75 \times \frac{9}{5})$ + 32 = 167°F.

To change from degrees Fahrenheit to degrees Centigrade, subtract 32 from the Fahrenheit temperature, and multiply by $\frac{5}{9}$.

Sample Problem 4

Water boils at 212° Fahrenheit. What is the equivalent temperature in degrees Centigrade?

$(212°F. - 32) \times \frac{5}{9}$ = 100 °C.

Problems	Work Here

(1) Change 15 miles to kilometers.

Answer:

(2) Eight liters is equivalent to how many U.S. gallons?

Answer:

(3) Change 9963.25 millimeters to feet and inches.

Answer:

(4) Find the millimeter equivalent of 75 feet $4\frac{13}{16}$ inches.

Answer:

(5) Find the millimeter equivalent of 143.66 feet.

Answer:

Problems	Work Here

(6-8) A room is 2971.8 millimeters by 3822.7 millimeters by 2438.4 millimeters high. Find its dimensions in feet and inches.

Answer:

Answer:

Answer:

(9) Change $99\frac{1}{2}°$ Fahrenheit to Centigrade temperature.

Answer:

(10) Find the Fahrenheit equivalent to 20° Centigrade.

Answer:

Powers and Square Roots

Study pages 249 to 262 of your textbook, <u>Practical</u> <u>Mathematics</u>, before preparing this test.

DO NOT SUBMIT THIS SELF-CHECK TEST FOR GRADING. Check your answers with the Answer Key at the end of this workbook.

Powers

Information

  Sometimes it is necessary for a carpenter to find missing dimensions in estimating for a given job. This usually involves the use of a formula. For example, $A = s^2$ is the formula which can be used to find the area (A) of a square when the side (s) is given. To use this formula requires an understanding of powers.

What Is a Power?

  Whenever two or more numbers are multiplied together to give a <u>product</u>, we refer to the separate numbers as <u>factors</u>. For example, $5 \times 3 \times 2 = 30$. In this problem 5, 3, and 2 are <u>factors of 30</u>.

  Now suppose these factors are all the same number such as $5 \times 5 \times 5$ or $3 \times 3 \times 3$ or $2 \times 2 \times 2$. The products of these numbers, 125, 27, and 8, are called <u>powers</u>. The factors 5, 3, and 2 are called the <u>bases</u> of the powers 125, 27, and 8. If a <u>base</u> is used as a <u>factor</u> a certain number of times, the resulting <u>product</u> is a <u>power</u>.

  A small number called an <u>exponent</u> is used to indicate the number of times the base is to be used as a factor. It is placed above and to the right of the base. For example, 7^2 means 7 <u>squared</u> or 7×7 or 49, 3^3 means 3 <u>cubed</u> or $3 \times 3 \times 3$ or 27, and 6^4 means 6 <u>to the fourth power</u> or $6 \times 6 \times 6 \times 6$ or 1,296.

Sample Formulas

    (1) Area of a square: $A = s^2$

    (2) Area of a circle: $A = .7854d^2$

    (3) Volume of a cube: $V = e^3$

    (4) Volume of a cylinder: $V = .7854d^2h$

Problems	Work Here

(1) Find the value of 18^2, or the square of 18.

Answer:

(2) What is the value of $(2.5)^3$, or the cube of 2.5?

Answer:

(3) Find the value of $(6.875)^2$.

Answer:

(4) What is the third power of 6.75?

Answer:

(5) Raise $\frac{3}{4}$ to the third power.

Answer:

Square Roots

Information

A carpenter, in finding missing dimensions, must sometimes find the square root of a number. To find the length (s) of one side of a square when the area (A) is given, the formula $s = \sqrt{A}$ is used. To use this formula requires an understanding of square roots.

What Is a Square Root?

The process of finding the number which when multiplied by itself equals a given number is called <u>extracting</u> the <u>square root</u>. Since we know that 5×5 is equal to 25, we say that 5 is the square root of 25.

We already know that 1, 2, 3, 4, 5, 6, 7, 8, 9, 10, 11, and 12 are the square roots of 1, 4, 9, 16, 25, 36, 49, 64, 81, 100, 121, and 144. To indicate that the square root of a number is to be found, we use the sign $\sqrt{}$

$$\sqrt{1} = 1 \qquad \sqrt{16} = 4 \qquad \sqrt{49} = 7 \qquad \sqrt{100} = 10$$

$$\sqrt{4} = 2 \qquad \sqrt{25} = 5 \qquad \sqrt{64} = 8 \qquad \sqrt{121} = 11$$

$$\sqrt{9} = 3 \qquad \sqrt{36} = 6 \qquad \sqrt{81} = 9 \qquad \sqrt{144} = 12$$

Sample Formulas

(1) Side of a square: $\qquad s = \sqrt{A}$

(2) Diameter of a circle: $\qquad d = \sqrt{\dfrac{A}{.7854}}$

(3) Diagonal of a rectangle: $\qquad d = \sqrt{l^2 + w^2}$

(4) Diameter of a cylinder: $\qquad d = \sqrt{\dfrac{V}{.7854h}}$

(5) Diameter of a sphere: $\qquad d = \sqrt{\dfrac{S}{3.1416}}$

Problems	Work Here

(1) Find the square root of 625.

Answer:

(2) Find the square root of 8,281.

Answer:

(3) Find the square root of 1.1449.

Answer:

(4) Find the square root of 126. Give
the answer in three decimal places.

Answer:

(5) Find the square root of 6.2480.

Answer:

Ratio

Study pages 267 to 278 of your textbook, <u>Practical</u> <u>Mathematics</u>, before preparing this test.

DO NOT SUBMIT THIS SELF-CHECK TEST FOR GRADING. Check your answers with the Answer Key at the end of this workbook.

Problems	Work Here

In the following problems state ratios as follows:

> 1 to 2 = 1:2
> 2 to 3 = 2:3
> 1 to 4 = 1:4, etc.

(1) Two boards are 4 in. and 6 in. wide. Find the ratio of their widths.

Answer:

(2) A house is 42 ft. long and 30 ft. wide. What is the ratio of the length to the width?

Answer:

(3) What is the ratio of the amounts of sand in two piles containing 70 cu. yd. and 105 cu. yd.?

Answer:

(4) Find the ratio of the lengths of two shelves 32 in. and 3 ft. 4 in. long.

Answer:

(5) What is the ratio of the width to the height of a door 2 ft. 8 in. by 6 ft. 8 in.?

Answer:

Problems	Work Here

(6-7) Find the ratio of the thickness to the
length, and find the ratio of the width
to the length of a standard size brick
$2\frac{1}{4}$ in. by $3\frac{3}{4}$ in. by 8 in.

Answer:

Answer:

(8) The circumferences of 15-inch and 20-
inch circular pieces of plywood are
47.124 in. and 62.832 in. Find the
ratio of the circumferences.

Answer:

(9-10) Find the ratio of the circumferences
and areas of 5-inch and 10-inch circular pieces of sheet metal if the
circumferences are 15.708 in. and
31.416 in. and the areas are 19.635 sq.
in. and 78.54 sq. in.

Answer:

Answer:

Proportion

Study pages 278 to 293 of your textbook, Practical Mathematics, before preparing this test.

DO NOT SUBMIT THIS SELF-CHECK TEST FOR GRADING. Check your answers with the Answer Key at the end of this workbook.

Problems	Work Here

Find the missing term in each of the following:

(1) 1:2 = 8 in.:?

Answer:

(2) 4:3 = ? : 22.5 ft.

Answer:

(3) 5:? = 45 yd. : 63 yd.

Answer:

(4) ? : 12.25 = 50 lb.: 61.25 lb.

Answer:

(5) $\frac{72}{48} = \frac{660 \text{ cu. yd.}}{?}$

Answer:

Problems	Work Here

(6) Find the number of pounds of nails
required for 3,750 lath if each
thousand lath requires 6 lb. of nails.

Answer:

(7) How many pounds of nails will be
required for 1,825 sq. ft. of metal
lath if 9 lbs. are required for each
1,000 sq. ft.?

Answer:

(8) For plaster work on wooden laths
about $1\frac{1}{2}$ cu. yd. of sand are needed
for each 100 sq. yd. How much
sand will be needed for 4,752 sq. ft.?

Answer:

(9) Determine the quantity of priming
paint needed for 3,150 sq. ft. if one
gallon covers 700 sq. ft.

Answer:

(10) If six square feet of 8-inch brick wall
with $\frac{1}{4}$-inch joints contains 78 bricks,
find the number of bricks needed for
125 sq. ft.

Answer:

Formulas—Meaning and Uses

Study pages 299 to 334 of your textbook, <u>Practical</u> <u>Mathematics</u>, before preparing this test.

DO NOT SUBMIT THIS SELF-CHECK TEST FOR GRADING. Check your answers with the Answer Key at the end of this workbook.

Information

A thorough understanding of the meaning and the use of formulas is essential in making estimates in the various trades.

What Is a Formula?

A <u>formula</u> is a rule expressed in simplified form by the use of <u>numbers</u>, <u>letters</u>, or <u>symbols.</u>

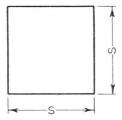

Rule: To find the area of a square, multiply the length of one side of the square by itself.

If we refer to Lesson 23 we can see that multiplying one side (s) by itself is the same as finding the square of s. The square of s is s² which is the same as s × s. If we represent the area by A, then the formula for the area of a square is $A = s^2$.

It is important in learning how to use formulas to realize that they can be written in other forms. Thus if we have the formula $A = s^2$, we can think of it in a new form as A = s × s.

Writing Formulas in a New Form

Formula	New Form

(1) Perimeter of a square is equal to four times one side:

$\underline{P = 4s}$ (1) $P = 4 \times s$

(2) Perimeter of a rectangle equals the sum of the sides or twice the length plus twice the width:

$\underline{P = 2b + 2h}$ (2) $P =$

(3) Circumference of a circle is equal to pi times the diameter:

$\underline{C = \pi d}$ (3) $C =$

(4) Circumference of a circle is equal to twice pi times the radius:

$\underline{C = 2\pi r}$ (4) $C =$

(5) Diagonal of a square equals 1.414 times one side:

$\underline{d = 1.414s}$ (5) $d =$

(6) Area of a square is equal to the square of the side:

$\underline{A = s^2}$ (6) $A = s \times s$

(7) Area of a rectangle is equal to the product of the base and the height:

$\underline{A = bh}$ (7) $A =$

(8) Area of a circle equals pi times the square of the radius:

$\underline{A = \pi r^2}$ (8) $A =$

(9) Area of a triangle is equal to $\frac{1}{2}$ the product of the base and the altitude:

$\underline{A = \frac{1}{2}bh}$ (9) $A =$

(10) Area of an equilateral triangle equals .433 times the square of one of the sides:

$\underline{A = .433s^2}$ (10) $A =$

(11) Area of a regular hexagon is equal to 2.598 times the square of the side:

$\underline{A = 2.598s^2}$ (11) $A =$

Writing Formulas in a New Form (continued)

Formula	New Form

(12) Lateral surface of a sphere is equal to four times the area of a great circle:

$S = 4\pi r^2$ (12) $S = 4 \times \pi \times r \times r$

(13) Volume of a cube equals the cube of an edge:

$V = e^3$ (13) $V =$

(14) Volume of a cylinder is found by multiplying the area of its base by its altitude:

$V = .7854d^2h$ (14) $V =$

(15) Volume of a sphere is equal to .5236 times the cube of the diameter:

$V = .5236d^3$ (15) $V =$

(16) Volume of a cone is the product of one-third its base area and its altitude:

$V = .2618d^2h$ (16) $V =$

(17) In a right triangle the side opposite the right angle is equal to the square root of the sum of the squares of the other two sides:

$c = \sqrt{a^2 + b^2}$ (17) $c = \sqrt{(a \times a) + (b \times b)}$

(18-19) In a right triangle, either of the sides forming the right angle may be found by subtracting the square of the other side from the square of the side opposite the right angle and taking the square root:

$a = \sqrt{c^2 - b^2}$ (18) $a =$

$b = \sqrt{c^2 - a^2}$ (19) $b =$

(20) Diameter of a cylinder is found by taking the square root of the volume divided by .7854 times the altitude:

$d = \sqrt{\dfrac{V}{.7854h}}$ (20) $d =$

Substituting Values in Formulas

Substitute the given values in the formulas but do not complete the work.

	Formulas	Given Values	Work Here
(1)	Perimeter of a square: $P = 4s$	$s = 8$	$P = 4 \times 8$
(2)	Perimeter of a rectangle: $P = 2l + 2w$	$l = 12, \ w = 7$	$P = (2 \times 12) + (2 \times 7)$
(3)	Circumference of a circle: $C = \pi d$	$\pi = 3.14, \ d = 10$	$C =$
(4)	Circumference of a circle: $C = 2\pi r$	$\pi = 3.14, \ r = 6.5$	$C =$
(5)	Diagonal of a square: $d = 1.414s$	$s = 12.75$	$d =$
(6)	Area of a square: $A = s^2$	$s = 8.875$	$A =$
(7)	Area of a rectangle $A = lw$	$l = 14.5, \ w = 6.25$	$A =$
(8)	Area of a triangle: $A = \frac{1}{2}bh$	$b = 24, \ h = 14$	$A =$
(9)	Area of an isoceles triangle: $A = .433s^2$	$s = 11$	$A =$
(10)	Area of a regular hexagon: $A = 2.598s^2$	$s = 15$	$A =$
(11)	Lateral surface of a cylinder: $S = 2\pi rh$	$\pi = 3.14, \ r = 8,$ $h = 5$	$S =$

Substituting Values in Formulas (continued)

Formulas	Given Values	Work Here

(12) Lateral surface of a sphere:

$S = 4\pi r^2$ $\quad\quad$ $\pi = 3.14, \ r = 7$ $\quad$ S =

(13) Area of a trapezoid:

$A = \dfrac{1}{2}h \ (b_1 + b_2)$ $\quad\quad$ $h = 10, \ b_1 = 12,$
$\quad\quad\quad\quad\quad\quad\quad\quad\quad b_2 = 16$ $\quad$ A =

(14) Volume of a cylinder:

$V = \pi r^2 h$ $\quad\quad$ $\pi = 3.14, \ r = 9,$
$\quad\quad\quad\quad\quad\quad\quad\quad h = 5.5$ $\quad$ V =

(15) Side of a square:

$s = \sqrt{A}$ $\quad\quad$ $A = 125$ $\quad$ $s = \sqrt{125}$

(16) Diameter of a circle:

$d = \sqrt{\dfrac{A}{.7854}}$ $\quad\quad$ $A = 628.32$ $\quad$ d =

(17) Diagonal of a rectangle:

$d = \sqrt{l^2 + w^2}$ $\quad\quad$ $l = 10, \ w = 6$ $\quad$ d =

(18) Diameter of a cylinder:

$d = \sqrt{\dfrac{V}{.7854h}}$ $\quad\quad$ $V = 431.97, \ h = 10$ $\quad$ d =

(19) Altitude of a trapezoid:

$h = \dfrac{2A}{b_1 + b_2}$ $\quad\quad$ $A = 288, \ b_1 - 18,$
$\quad\quad\quad\quad\quad\quad\quad\quad b_2 = 24$ $\quad$ h =

(20) Side of a triangle:

$a = \dfrac{2A}{b}$ $\quad\quad$ $A = 75, \ b = 18.75$ $\quad$ a =

Sample Problem 1

Find the area of a square 8 in. on each side.

Instructions	Computations
(1) Write the formula.	(1) $A = s^2$
(2) Write the formula in a new form	(2) $A = s \times s$
(3) Write the given value.	(3) $s = 8$
(4) Substitute the value in the formula.	(4) $A = 8 \times 8$
(5) Multiply.	(5) $A = 64$

Answer: Area = 64 sq. in.

Sample Problem 2

What is the area of a circle whose diameter is 15 in. ?

Instructions	Computations
(1) Write the formula.	(1) $A = .7854d^2$
(2) Write the formula in a new form.	(2) $A = .7854 \times d \times d$
(3) Write the given value.	(3) $d = 15$
(4) Substitute this value in the formula.	(4) $A = .7854 \times 15 \times 15$
(5) Multiply.	(5) $A = 176.715$

Answer: Area = 176.715 sq. in.

Sample Problem 3

The edges of a cube are 5.5 in. long. Find its volume.

Instructions	Computations
(1) Write the formula.	(1) $V = e^3$
(2) Write the formula in a new form.	(2) $V = e \times e \times e$
(3) Write the given value.	(3) $e = 5.5$
(4) Substitute this value in the formula.	(4) $V = 5.5 \times 5.5 \times 5.5$
(5) Multiply.	(5) $V = 166.375$

Answer: Volume = 166.375 cu. in.

Sample Problem 4

Find the length of the diagonal of a rectangle which is 24 in. long and 18 in. wide.

Instructions	Computations
(1) Write the formula	(1) $d = \sqrt{l^2 + w^2}$
(2) Write the formula in a new form.	(2) $d = \sqrt{(l \times l) + (w \times w)}$
(3) Write the given values.	(3) $l = 24$, $w = 18$
(4) Substitute these values in the formula.	(4) $d = \sqrt{(24 \times 24) + (18 \times 18)}$
(5) Multiply.	(5) $d = \sqrt{576 + 324}$
(6) Add.	(6) $d = \sqrt{900}$
(7) Find the square root.	(7) $d = 30$

Answer: $d = 30$ in.

Problems	Work Here

(1) The length of a side of a square is 14 in. What is the area of the square? (Area of a square equals the side squared.)

Answer:

(2) Find the area of a circle whose diameter is 18 in. (Area of a circle equals π times the radius squared.)

Answer:

Problems	Work Here

(3) What is the volume of a cylinder
whose diameter is 5 in. and height
8 in. ? (The volume of a cylinder
is equal to .7854 times the diameter
squared times the height.)

Answer:

(4) A rectangle is 18 in. long and 13.5
in. wide. What is the length of its
diagonal? (The square of the diagonal
is the sum of the squares of the sides.)

Answer:

(5) The volume of a cylinder is 500 cu. in.
Find its diameter if its height is 15 in.
(The volume of a cylinder is equal to
.7854 times the diameter squared times
the height.)

Answer:

RELATED MATHEMATICS FOR CARPENTERS

Examination No. 3

Based on pages 47 to 74 in this workbook

Student's Name_____ Student Number_____

Street_____City_____State_____Code_____ Zip

Problems	Show Work Here

(1) Find 65 per cent of 1,875 ft.

Answer:

(2) What per cent of $\frac{15}{16}$ of an inch is $\frac{3}{8}$ of an inch?

Answer:

(3) Find the per cent of finish floor-ing which <u>remains</u> to be applied after 1,825 ft. of 2,700 ft. were put on the subflooring.

Answer:

(4) How many square feet of plywood are there in a piece containing 2,664 sq. in.?

Answer:

(5) How many linear feet of sheathing are there in 1,275 pieces each $11\frac{1}{2}$ ft. long?

Answer:

Problems	Show Work Here

(6) Change 3 yd. 1 ft. 5 in. to inches.

Answer:

(7) Subtract 4 hr. 25 min. from 7 hr. 20 min.

Answer:

(8) What is the total length of three rafters 12 ft. 5 in., 8 ft. 8 in., and 5 ft. 11 in. long?

Answer:

(9) Find the difference in length of two pieces of flooring 7 ft. 2 in. and 5 ft. 8 in. long.

Answer:

(10) How long will it take to complete 16 jobs, each requiring 3 hr. 35 min.?

Answer:

(11) Find the total area in square feet of three boards whose dimensions are 4 in. by 12 ft., 8 in. by 18 ft., and 12 in. by 16 ft.

Answer:

Student's Name_____Student Number_____

Problems	Show Work Here

(12-13) The circumference of a 5-inch
circular piece of plywood is
15.708 in. and that of a 15-inch
circular piece is 47.124 in. What
is the ratio of the diameters and
circumferences?

Answer:

Answer:

(14-15) The area of a 6-inch circle is
28.2744 sq. in. and that of a 12-
inch circle 113.0976 sq. in. What
is the ratio of the diameters?
What is the ratio of the areas?

Answer:

Answer:

(16) Using the formula d $= \sqrt{l^2 + w^2}$,
find d if l = 72 and w = 54.

Answer:

Problems	Show Work Here

(17) Using the formula $V = .7854d^2 h$,
      find the volume of a cylinder with
      a 7-inch diameter and a height of
      12 inches.

Answer:

(18) Using the formula $d = \sqrt{\dfrac{A}{.7854}}$,
      find d if A = 84.8232.

Answer:

(19) Find the millimeter equivalent
      of 865 ft. $10\frac{7}{8}$ in.

Answer:

(20) Change 23,361.650 mm. to feet
      and inches.

Answer:

Squares

Study pages 384 to 385 and 397 to 398 of your textbook, <u>Practical</u> <u>Mathematics</u>, before preparing this test.

DO NOT SUBMIT THIS SELF–CHECK TEST FOR GRADING. Check your answers with the Answer Key at the end of this workbook.

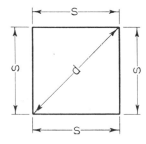

Information

        The <u>perimeter</u> (P) of a square is equal to four times the length of one side of the square.

        The diagonal (d) of a square is equal to the product of 1.414 and the length of one side.

        The area (A) of a square is equal to the product of its length and width or the square of its side s.

Formulas

$$(1)\ P = 4 \times s\ \ or\ \ P = 4s$$

$$(2)\ d = 1.414 \times s\ or\ d = 1.414s$$

$$(3)\ A = s \times s\ \ or\ \ A = s^2$$

Sample Problems

    (1) Find the perimeter of a square room whose sides are 12 ft. long.

$$P = 4 \times s = 4 \times 12 = 48$$

        The perimeter of the room is 48 ft.

    (2) What is the diagonal distance from any corner to the opposite corner of the ceiling of a square room whose sides are 10 ft. long?

$$d = 1.414 \times s = 1.414 \times 10 = 14.14$$

        The diagonal distance is 14.14 ft.   .14 ft. = .14 × 12 in. or 1.68 in.

.68 in. = $\frac{11}{16}$ in.(approximately). 14.14 ft. = 14 ft. $1\frac{11}{16}$ in.

(3) Determine the area of the floor of a square room whose sides are 15 ft. long.

$$A = s \times s = 15 \times 15 = 225$$

The area of the floor is 225 sq. ft.

Problems	Work Here

(1-3) Find the perimeter, area, and diagonal distance of a square lot 202 ft. on each side.

Answer:

Answer:

Answer:

(4) How many square feet of plywood will be required to cover the tops of 125 square tables each 3 ft. 6 in. on a side?

Answer:

(5) Find the approximate number of 3 in. by 3 in. tiles required for a kitchen floor 11 ft. wide and 13 ft. 3 in. long.

Answer:

(6) The floor of a locker room 25 ft. wide and 100 ft. long is covered with $4\frac{1}{4}$ in. by $4\frac{1}{4}$ in. tile. Determine the number of tiles required.

Answer:

Rectangles

Study pages 385 and 397 to 398 of your textbook, <u>Practical</u> <u>Mathematics</u>, before preparing this test.

DO NOT SUBMIT THIS SELF-CHECK TEST FOR GRADING. Check your answers with the Answer Key at the end of this workbook.

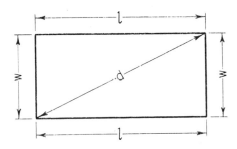

Information

The <u>perimeter</u> (P) of a rectangle is equal to twice the sum of its length and width.

The <u>diagonal</u> (d) of a rectangle is equal to the square root of the sum of the squares of its length and width.

The area (A) of a rectangle is equal to the product of its length and width.

Formulas

(1) $P = 2l + 2w$ or $P = 2(l + w)$ or $P = 2 \times (l + w)$

(2) $d = \sqrt{l^2 + w^2}$

(3) $A = l \times w$ or $A = lw$

Sample Problems

(1) Find the perimeter of a room if its length is 20 ft. and width 12 ft.

$$P = 2 \times (l + w) = 2 \times (20 + 12) = 2 \times 32 = 64$$

The perimeter of the room is 64 ft.

(2) What is the diagonal distance of a floor 20 ft. long and 15 ft. wide?

$$d = \sqrt{l^2 + w^2} = \sqrt{400 + 225} = \sqrt{625} = 25$$

The diagonal distance is 25 ft.

(3) Determine the area of a wall 25 ft. long and 12 ft. high.

$$A = l \times w = 25 \times 12 = 300$$

The area of the wall is 300 sq. ft.

Problems	Work Here

(1-3) Find the perimeter, area, and diagonal distance across a rectangular plot of ground 96 ft. wide and 128 ft. long.

Answer:

Answer:

Answer:

(4) Approximately how many oblong tiles 6 in. by $4\frac{1}{4}$ in. will be required to cover the floor of a shower room 15 ft. wide and 32 ft. long?

Answer:

(5-7) A room 24 ft. long, 16 ft. wide, and 9 ft. high has a door opening 6 ft. wide and 7 ft. high, and 6 windows 3 ft. 6 in. wide and 6 ft. high. Find the total area of the walls and ceilings without the openings, the total area of the openings, and the area less the openings.

Answer:

Answer:

Answer:

Parallelograms

Study pages 385 and 397 to 398 of your textbook, <u>Practical</u> <u>Mathematics</u>, before preparing this test.

DO NOT SUBMIT THIS SELF-CHECK TEST FOR GRADING. Check your answers with the Answer Key at the end of this workbook.

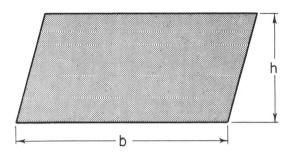

Information

    A <u>parallelogram</u> is a figure of four sides whose opposite sides are parallel and equal.

    The side on which the parallelogram rests is called the <u>base</u> (b).

    The shortest distance between two of the parallel sides is called the <u>altitude</u> (h).

Formula

$$A = bh \quad \text{or} \quad A = b \times h$$

Sample Problem

    Find the area of a parallelogram whose base is 18 in. and altitude 12 in.

$$A = b \times h$$
$$= 18 \times 12$$
$$= 216$$

    The area of the parallelogram is 216 sq. in.

Problems

    Find the area of each parallelogram.

	b	h	A
(1)	12 in.	8 in.	
(2)	2.5 ft.	14 in.	
(3)	18.25 in.	7.25 in.	
(4)	5 ft. 6 in.	2.75 ft.	
(5)	5.625 in.	3.875 in.	

Problems	Work Here

(6-7) Find the area of the top and front of a piece of bridging 20 in. long, 3 in. wide, and $1\frac{1}{2}$ in. thick.

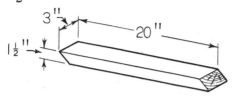

Answer:

Answer:

(8-9) What are the top and front areas of a brace 22 in. long, 6 in. wide, and 2 in. thick?

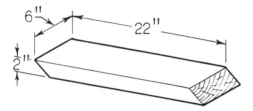

Answer:

Answer:

(10) A sidewalk 80 ft. long and 4 ft. wide cuts across a lawn as shown in this figure. By how many sq. ft. is the lawn area reduced by the sidewalk?

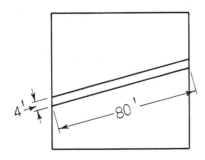

Answer:

Trapezoids

Study pages 385 and 399 to 400 of your textbook, <u>Practical Mathematics</u>, before preparing this test.

DO NOT SUBMIT THIS SELF-CHECK TEST FOR GRADING. Check your answers with the Answer Key at the end of this workbook.

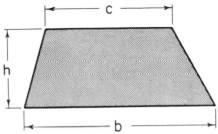

Information

A figure of four sides with only one pair of parallel sides is called a <u>trapezoid</u>.

If the sides which are not parallel are equal it is called an <u>isosceles trapezoid</u>.

The area of the trapezoid is equal to the product of one-half its altitude and the sum of its bases.

Formula

$$A = \frac{h}{2}(b + c) \text{ or } A = \frac{h(b + c)}{2} \text{ or } A = \frac{h \times (b + c)}{2}$$

Sample Problem

What is the area of a floor with parallel sides of 10 ft. and 6 ft. if it is 4 ft. wide?

$$A = \frac{h \times (b + c)}{2}$$

$$= \frac{\overset{2}{\cancel{4}} \times (10 + 6)}{\underset{1}{\cancel{2}}}$$

$$= 2 \times 16$$

$$= 32$$

The area of the trapezoid is 32 sq. ft.

Problems	Work Here

(1) Find the amount of floor space for a bay window if the parallel sides are 10 ft. and 14.5 ft. long and the distance between them 4 ft.

Answer:

Problems	Work Here

(2) What is the area of a shelf whose parallel sides are 34 in. and 51 in. if the width is 14 in.?

Answer:

(3) What is the area of the end of the trough?

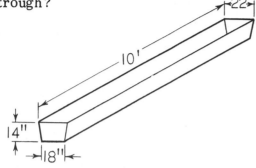

Answer:

(4) Find the area of this lot.

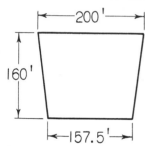

Answer:

(5) Determine the area of this sidewalk.

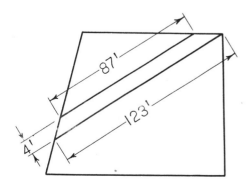

Answer:

Regular Hexagons

Study pages 386 and 402 to 404 of your textbook, <u>Practical Mathematics</u>, before preparing this test.

DO NOT SUBMIT THIS SELF–CHECK TEST FOR GRADING. Check your answers with the Answer Key at the end of this workbook.

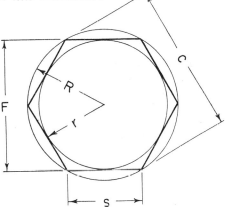

Information

A polygon having six equal sides is called a <u>regular hexagon.</u>

R = radius of circumscribed circle   A = area
r = radius of inscribed circle   P = perimeter

      S = length of one side. It is equal to R.
      C = distance across corners
      F = distance across flats

Formulas

(1) $r = .866S$ or $r = .866R$
(2) $R = 1.155r$ or $R = S$
(3) $A = 2.598S^2$ or $A = 2.598R^2$ or $A = 3.464r^2$
(4) $P = 6S$
(5) $C = 2S$ or $C = 2R$
(6) $F = 2r$ or $F = 1.732S$ or $F = 1.732R$

Sample Problems

Find r, R, A, P, C, and F for a regular hexagon if S = 8 in.

$r = .866 \times 8$ $r = 6.928$
$R = 8$ in.
$A = 2.598 \times 64$ $A = 166.072$ sq. in.
$P = 6 \times 8$ $P = 48$ in.
$C = 2 \times 8$ $C = 16$ in.
$F = 1.732 \times 8$ $F = 13.856$ in.

Problems	Work Here

(1) Find the area of the bottom of a
hexagonal wastebasket if the
edges are 6 in. long.

Answer:

(2-4) Find the distance in feet across corners of a regular hexagonal table top
with 30 in. edges. Find also the
amount of linoleum in cubic feet to
cover the top, and the length in feet
of aluminum edging.

Answer:

Answer:

Answer:

(5) Approximately how many hexagonal
tiles 2 in. on each side will be required for a floor 100 ft. long and
38 ft. wide.

Answer:

Triangles

Study pages 387 to 389 and 393 to 397 of your textbook, <u>Practical Mathematics</u>, before preparing this test.

DO NOT SUBMIT THIS SELF–CHECK TEST FOR GRADING. Check your answers with the Answer Key at the end of this workbook.

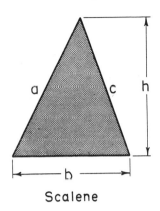

Scalene

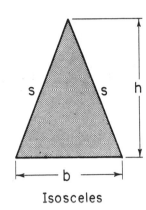

Isosceles

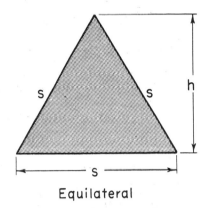

Equilateral

Information

A figure formed by three straight lines meeting at three points is called a <u>triangle</u>.

The points where the lines meet are called <u>vertices</u>.

Each point is a <u>vertex</u>.

The lines between these points are the <u>sides</u>.

A triangle with no two sides equal is a <u>scalene triangle</u>.
A triangle with two equal sides is an <u>isosceles triangle</u>.
A triangle with three equal sides is an <u>equilateral triangle</u>.

The sum of the lengths of the sides of a triangle is called its <u>perimeter</u>.

The <u>area</u> of a triangle is equal to one-half the product of its base (the side upon which it rests) and its altitude or height.

Formulas

(1) Scalene triangle:

$$P = a + b + c \qquad\qquad A = \tfrac{1}{2}hb \text{ or } A = \frac{h \times b}{2}$$

(2) Isosceles triangle:

$$P = s + s + b \qquad\qquad A = \tfrac{1}{2}hb \text{ or } A = \frac{h \times b}{2}$$
$$= 2s + b$$
$$= (2 \times s) + b$$

(3) Equilateral triangle:

$$P = s + s + s \text{ or } P = 3 \times s \qquad A = .433s^2 \text{ or } A = .433 \times s \times s$$

When no altitude is given but all three sides of a triangle are known, the area of a triangle may be determined by using the following formula:

$$\text{Area} = \sqrt{s\ (s - a)\ (s - b)\ (s - c)}$$

Where $s = \dfrac{1}{2}$ the perimeter.

Sample Problems

(1) What is the perimeter of a triangular plot of ground where sides are 100 ft., 80 ft., and 60 ft. long?

$$P = a + b + c$$
$$= 100 + 80 + 60$$
$$= 240$$

The perimeter of the plot of ground is 240 ft.

(2) Find the area of a triangular plot of ground whose base is 212 ft. and altitude 110 ft.

$$A = \frac{h \times b}{2}$$
$$= \frac{\overset{55}{\cancel{110}} \times 212}{\underset{1}{\cancel{2}}}$$
$$= 11,660$$

The area of the plot of ground is 11,660 sq. ft.

(3) The equal sides of an isosceles triangle are 20 in. long and its base 14 in. long. Find its perimeter.

$$P = (2 \times s) + b$$
$$= (2 \times 20) + 14$$
$$= 40 + 14$$
$$= 54$$

The perimeter of the isoceles triangle is 54 in.

(4) What is the area of a triangular piece of plywood whose sides are 22 in. long?

$$A = .433 \times s \times s$$
$$= .433 \times 22 \times 22$$
$$= .433 \times 484$$
$$= 209.572$$

The area of the piece of plywood is 209.572 sq. in.

(5) Find the area of a triangular plot whose sides are 6 ft., 8 ft. and 10 ft.

$$\text{Area} = \sqrt{12\ (12 - 10)\ (12 - 8)\ (12 - 6)}$$

$$= \sqrt{12 \times 2 \times 4 \times 6}$$

$$= \sqrt{576}$$

$$= 24 \text{ sq. ft.}$$

Problems	Work Here

(1) Find the number of square feet of area in the gable end of a garage if the base is 24 ft. and the height 10.5 ft.

Answer:

(2) What is the area of a triangular plot of ground if each of the sides is 150 ft. long?

Answer:

(3) Determine the number of square feet of plywood needed to cut 1,000 right triangles with bases 12 in. long and altitudes of 10 in.

Answer:

(4) How many triangular shaped tiles each 3 in. on a side will be needed to cover a floor 12 ft. long and 6 ft. wide?

Answer:

(5) What is the area of a triangular plot of ground whose borders measure 12 ft., 16 ft., and 20 ft.?

Answer:

SELF-CHECK TEST 33

Missing Dimensions and Irregular Figures

DO NOT SUBMIT THIS SELF-CHECK TEST FOR GRADING. Check your answers with the Answer Key at the end of this workbook.

Information

If several dimensions of a figure are known, often missing dimensions may be calculated.

Sample Problem

Find the inner width of a wall whose outside width is 16 feet 6 inches with 6-inch-thick walls.

First draw a picture to help visualize the information given:

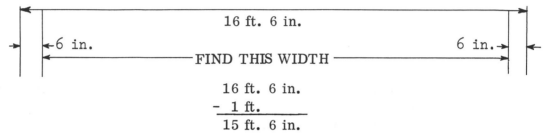

$$\begin{array}{r} 16 \text{ ft. } 6 \text{ in.} \\ - \ 1 \text{ ft.} \\ \hline 15 \text{ ft. } 6 \text{ in.} \end{array}$$

To find the area of an irregular shape, the usual procedure is to divide the area into smaller areas of common shapes, calculate the areas and then add them together.

Sample Problem

Find the total floor area of this house:

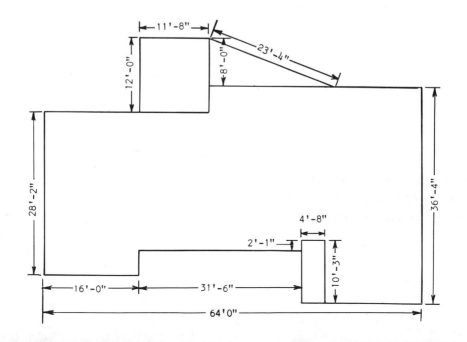

First, divide the area into smaller common-shaped figures.

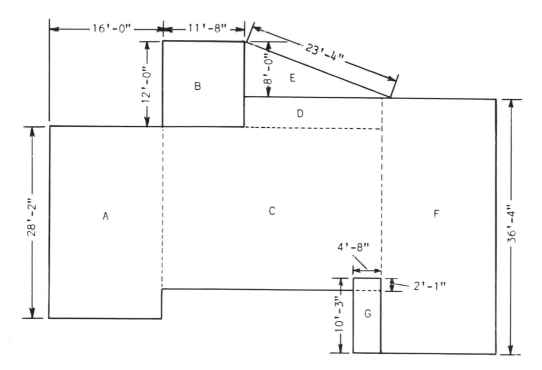

Next, find any dimensions still missing:

Find

Height of D   12'-0''   (B) − 8'-0''   (E) =   4'-0''

Width of D   31'-6''   (C) − 11'-8''   (B) = 19'-10''

Height of G to dotted line   10'-3''   (G) −   2'-1'' = 8'-2''

Height of C   36'-4''   (F) − 4'-0''   (D) − 8'-2''   (G) = 24'-2''

Base of E $(23'-4'')^2 - (8'-0'')^2 = 21.91'$

Figure each area separately.

A = 28'-2'' × 16'-0''   =   450.56 sq. ft.

B = 12'-0'' × 11'-8''   =   139.92 sq. ft.

C = 31'-6'' × 24'-2''   =   761.23 sq. ft.

D = 4'-0'' × 19'-10''   =   68.33 sq. ft.

E = $\frac{1}{2}$ × 21.91' × 8'   =   11.64 sq. ft.

F = 36'-4'' × 16'-6''   =   599.44 sq. ft.

G = 8'-2'' × 4'-8''   =   38.05 sq. ft.

Total   2460.86 sq. ft. = 2461 sq. ft.

Problems	Work Here

(1-2) Find missing dimensions A and
B in this figure:

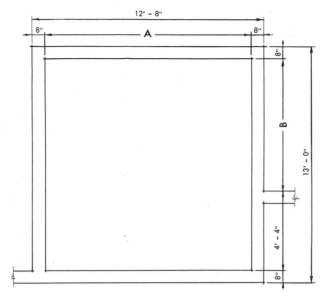

Answer:

Answer:

(3-7) Find the area of each section
and the total area of this
irregular figure:

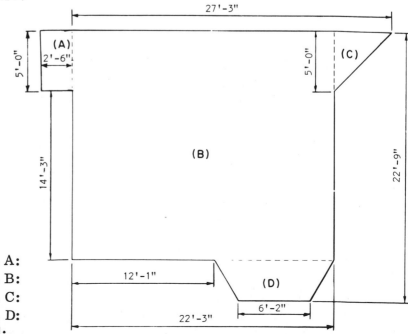

Answer Area A:
Answer Area B:
Answer Area C:
Answer Area D:
Answer Total:

RELATED MATHEMATICS FOR CARPENTERS

Examination No. 4

Based on Pages 79 to 94 in this workbook

Student's Name_____ Student Number_____

Street_____ City_____ State_____ Zip Code_____

Problems	Show Work Here

(1-3) Find the perimeter, length of diagonal, and area of a square piece of wallboard having 22-inch sides.

Answer:

Answer:

Answer:

(4-5) What is the length of the diagonal and area of a piece of lumber whose length is 2 ft. 3 in. and width 12 in.?

Answer:

Answer:

Problems	Show Work Here

(6) The parallel sides of a piece of
    sheathing are 28 in. and 34 in.
    long. It is 6 in. wide. What is
    its area?

Answer:

(7) Find the area of a gable 28.5 ft.
    long and 12 ft. from the ridge to
    the plate.

Answer:

(8) What is the area of a triangular
    plot of ground 80 ft. on each side?

Answer:

(9) What is the area of a triangular
    section of floor whose edges are
    3 feet, 5 feet and 4 feet?

Answer:

(10) The edges of a hexagonal
     table top are 9 in. long.
     What is the area?

Answer:

Circles

Study pages 405 to 410 of your textbook, <u>Practical</u> <u>Mathematics</u>, before preparing this test.

DO NOT SUBMIT THIS SELF-CHECK TEST FOR GRADING. Check your answers with the Answer Key at the end of this workbook.

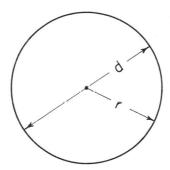

Information

A closed curve such that all points on it are at the same distance from a point inside called the center is a <u>circle</u>.

A straight line connecting two points on the circle and passing through its center is called a <u>diameter</u>.

A straight line joining the center with any point on the circle is called a <u>radius</u>.

The distance around the circle is its <u>circumference</u>.

Formulas

(1) $C = 2\pi r$ or πd $C = \pi \times d$ or $3.1416 \times d$

(2) $A = \dfrac{\pi d^2}{4}$ or $.7854 d^2$ $A = .7854 \times d \times d$

(3) $d = \dfrac{C}{\pi}$ or $\dfrac{C}{3.1416}$

(4) $d = \sqrt{\dfrac{A}{.7854}}$

Sample Problems

(1) Find the circumference of a circle whose diameter is 18 in.

$$C = 3.1416 \times d$$
$$= 3.1416 \times 18$$
$$= 56.5488$$

The circumference of the circle is 56.5488 in.

(2) Find the area of a circle whose diameter is 16 in.

$$A = .7854 \times d \times d$$
$$= .7854 \times 16 \times 16$$
$$= .7854 \times 256$$
$$= 201.0624$$

The area of the circle is 201.0624 sq. in.

(3) Find the diameter of a circle whose circumference is 72 in.

$$d = \frac{C}{3.1416}$$
$$= \frac{72}{3.1416}$$
$$= 22.9$$

The diameter of the circle is 22.9 in.

(4) Find the diameter of a circle whose area is 188 sq. in.

$$d = \sqrt{\frac{A}{.7854}}$$
$$= \sqrt{\frac{188}{.7854}}$$
$$= \sqrt{239.3685}$$
$$= 15.47$$

The diameter of the circle is 15.47 in.

Problems	Work Here

(1) What is the circumference of a circle whose diameter is 15 in.?

Answer:

(2) Find the area of a 22-in. circle.

Answer:

(3) What is the diameter of a circle whose circumference is 32 in.?

Answer:

(4) Find the diameter of a circle whose area is 156 sq. in.

Answer:

(5) Find the area of a circular table top if its diameter is 4 ft.

Answer:

(6) What is the area of the bottom of a silo if its diameter is 22 ft.?

Answer:

Problems	Work Here

(7) Determine the area of the bottom of a wooden vat if its diameter is 18.5 ft.

Answer:

(8) Find the total area of 135 wooden disks each 9 in. in diameter.

Answer:

(9) What is the area of a circular grill $21\frac{3}{4}$ in. in diameter?

Answer:

Ellipses

DO NOT SUBMIT THIS SELF-CHECK TEST FOR GRADING. Check your answers with the Answer Key at the end of this workbook.

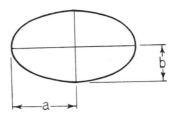

Information

The ellipse is very much like the circle.

The long diameter is called the <u>major axis</u> and the short diameter the <u>minor axis</u>.

Formulas

(1) $p = 3.1416 \sqrt{2(a^2 + b^2)} = 3.1416 \times \sqrt{2 \times (a^2 + b^2)}$

(2) $A = 3.1416 \times a \times b$ where $a = \frac{1}{2}$ of the major (long) axis and

$b = \frac{1}{2}$ of the minor (short) axis.

Sample Problems

(1) The length of the inside curve of an arch is one-half the periphery, or distance around an ellipse. What is this length if the distance between supports is 10 ft. and the height of the arch 4 ft. ?

$$\frac{1}{2} p = \frac{3.1416 \times \sqrt{2 \times (25 + 16)}}{2}$$

$$= 1.5708 \times \sqrt{2 \times 41}$$

$$= 1.5708 \times \sqrt{82}$$

$$= 1.5708 \times 9.055$$

$$= 14.22359 \text{ or } 14.224$$

The approximate length is 14 ft. $2\frac{11}{16}$ in.

(2) How many square feet of floor are there in a bay if the shape is one-half of an ellipse, (semielliptical) the length is 20 ft., and the depth 5 ft.?

$$\frac{1}{2}A = \frac{3.1416 \times 10 \times 5}{2}$$

$$= 78.54$$

The area is approximately 78 1/2 sq. ft.

Problems	Work Here

(1) The distance between supports of a
semielliptical arch is 12 ft. Find
the length of the inside curved surface
if it is 2.5 ft. high.

Answer:

(2) How many square feet of plywood are
required for the inside surface of a
semielliptical arch if the distance between supports is 18 ft., the height
4 ft., and the thickness of the wall 18 in.?

Answer:

(3) How many square feet of lumber are
necessary to replace the bottom of an
elliptical vat whose diameters are
30 ft. and 16 ft.?

Answer:

(4) How many square feet of material are
required for a form in the shape of one-
half an ellipse if the diameters are 6 ft.
and 2 ft., and the height 5 ft.?

Answer:

Circular Rings

Study pages 410 to 411 of your textbook, <u>Practical</u> <u>Mathematics</u>, before preparing this test.

DO NOT SUBMIT THIS SELF–CHECK TEST FOR GRADING. Check your answers with the Answer Key at the end of this workbook.

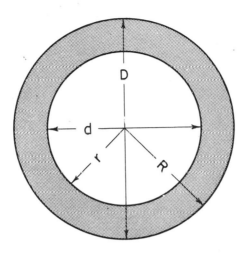

Information

A circular ring has an inside and an outside radius and an inside and an outside diameter.

The area is the number of square units contained in the ring.

Formulas　　　　(1) $A = \pi(R^2 - r^2)$ or $3.1416 \times (R^2 - r^2)$

(2) $A = \pi \dfrac{(D^2 - d^2)}{4}$ or $.7854 \times (D^2 - d^2)$

Sample Problems

(1) Find the area of a circular ring whose radii are 12 in. and 7 in.

$$A = 3.1416 \times (144 - 49)$$
$$= 3.1416 \times 95$$
$$= 298.452$$

The area of the circular ring is 298.45 sq. in. or 298 1/2 sq. in.

(2) Find the area of a circular ring whose diameters are 24 in. and 14 in.

$$A = .7854 \times (576 - 196)$$
$$= .7854 \times 380$$
$$- 298.452$$

The area of the circular ring is 298.45 sq. in. or 298 1/2 sq. in.

Problems	Work Here

(1) Find the area of a circular ring
if the radii are 9 in. and 11 in.

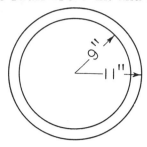

Answer:

(2) What is the area of a sidewalk 4 ft.
wide around a circular plot of
ground whose diameter is 76 ft.?

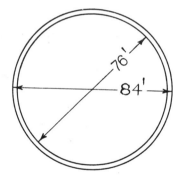

Answer:

(3) The diameter of a circular piece of
sheet metal is 25 in. A circular
piece 10 in. in diameter is cut from
the center. What is the area of the
ring?

Answer:

Radii, Diameters, and Chords of Circles

DO NOT SUBMIT THIS SELF-CHECK TEST FOR GRADING. Check your answers with the Answer Key at the end of this workbook.

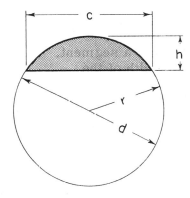

Information

It is sometimes necessary to determine the radius or diameter of a circle when the length of a chord and the height of the segment of a chord are given.

$$d = \text{diameter of circle} \qquad h = \text{height of segment}$$
$$r = \text{radius of circle} \qquad c = \text{length of chord}$$

Formulas

$$(1) \quad r = \frac{c^2 + 4h^2}{8h} = \frac{(c \times c) + (4 \times h \times h)}{8 \times h}$$

$$(2) \quad d = \frac{c^2 + 4h^2}{4h} = \frac{(c \times c) + (4 \times h \times h)}{4 \times h}$$

Sample Problem

The distance between the supports of a circular arch is 4 ft. 4 in. The height of the arch is 20 in. What is the diameter of the arch?

$$d = \frac{(52 \times 52) + (4 \times 20 \times 20)}{4 \times 20}$$

$$= \frac{2704 + 1600}{80}$$

$$= \frac{4304}{80}$$

$$= 53.8$$

The diameter of the arch is 53.8 in. or approximately 4 ft. $5\frac{13}{16}$ in.

Problems	Work Here

(1) A cabinetmaker found it necessary
to make a circular table top similar
to one which had been broken. He
had only a portion of the original
top which was in the form of a segment.
The greatest length and width of the
piece were 26 in. and 3 in. Find the
diameter of the original top.

Answer:

(2) A portion of a wooden pulley in the
form of a segment is to be replaced.
What is the radius of the new pulley
if the length of the chord is 8 in. and the
height of the segment 2.5 in.?

Answer:

(3) What is the radius of a circular bay
whose greatest length is 10 ft. and
greatest depth 3.5 ft.?

Answer:

Arcs and Sectors of Circles

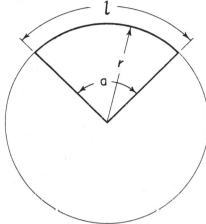

Information

An arc is a part of the circumference of a circle.

The part of a circle bounded by two radii and the included arc of a circle is called a sector.

r = radius of circle a = angle in degrees
l = length of arc A = area

Formulas

(1) $l = \dfrac{1.5708ra}{90°} = \dfrac{1.5708 \times r \times a}{90°}$

(2) $A = \dfrac{1}{2}r\,l = \dfrac{1}{2} \times r \times l = \dfrac{r \times l}{2}$

Sample Problems

(1) Find the length of the curved surface of an arch if the angle contains 72° and the radius is 5 ft.

$$l = \frac{1.5708 \times \overset{1}{\cancel{5}} \times \overset{4}{\cancel{72}}}{\underset{\underset{1}{\cancel{5}}}{90}} = 6.2832 \text{ ft. or 6 ft. } 3\frac{25}{64} \text{ in.}$$

The length of the curved surface of the arch is approximate · 6 ft. $3\frac{3}{8}$ in.

(2) A portion of the bottom of a cylindrical wooden vat 22 ft. in diameter is to be replaced. How many square feet of lumber will be required if the angle measures approximately 81°?

$$l = \frac{1.5708 \times 11 \times \overset{9}{\cancel{81}}}{\underset{10}{\cancel{90}}} = \frac{155.5092}{10} = 15.55 \text{ or } 15.55 \text{ ft.}$$

$$A = \frac{11 \times 15.55}{2} = \frac{171.05}{2} = 85.525$$

Approximately 85 1/2 sq. ft. will be required.

Problems	Work Here

(1) The angle of an arch contains 67.5°.
What is the inside length of the
curved surface if its radius is 25 ft.?

Answer:

(2) How many square feet of plywood
will be required to cover the inside
surface of an arch if the radius is
10 ft., the angle contains 75°, and the
wall is 15 in. thick?

Answer:

(3) How many square feet of material are
required to construct the outside arc
form for a bay if the radius is 15 ft.,
the angle 63°, and the height of the
form 6 ft. ?

Answer:

(4) Part of the flooring in a circular arena
200 ft. in diameter was damaged by
fire. Find the number of square feet
to replace the damaged sector if the
angle is approximately 60°.

Answer:

RELATED MATHEMATICS FOR CARPENTERS

Examination No. 5

Based on Pages 97 to 108 in this workbook

Student's Name_____Student Number_____

Street_____City_____State_____Zip Code_____

Problems	Show Work Here

(1-2) What is the circumference and area of a circular table top if its diameter is 5 ft.?

Answer:

Answer:

(3)   Find the area of the bottom of a round wooden tank whose diameter is 22.5 ft.

Answer:

(4)   A circle 8 inches in diameter is cut out of a circular piece of plywood which contains 100 square inches. What is the area of the remaining ring?

Answer:

(5)   The distance between supports of a semielliptical arch is 10 ft. and the height is 2 ft. The thickness of the wall is 16 in. How many square feet of plywood are necessary to cover the inside surface?

Answer:

Problems	Show Work Here

(6) Find the area of a sidewalk in the
shape of a circular ring if the inside and outside diameters are 56
ft. and 60 ft.

Answer:

(7) What is the area of the bottom of a
water tank if it is elliptical in shape
with diameters of 12 ft. and 8 ft.?

Answer:

(8) Find the circumference of a pond
whose diameter is 20 feet.

Answer:

(9) The angle of an arc contains 72°.
Find the length of the inside curve
of the arc if the radius is 18 ft.

Answer:

(10) A sector of a wheel 32 inches
in diameter is to be replaced.
Find the area if its angle is 60°.

Answer:

Surfaces of Cubes and Rectangular Solids

Study page 422 of your textbook, <u>Practical</u> <u>Mathematics</u>, before preparing this test.

DO NOT SUBMIT THIS SELF–CHECK TEST FOR GRADING. Check your answers with the Answer Key at the end of this workbook.

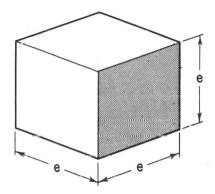

Surfaces of Cubes

Information

The sides of the cube are called the <u>lateral faces.</u>
The lateral surface of a cube is equal to the area of one face multiplied by 4.
The total surface of a cube is equal to the area of one face multiplied by 6.

Formulas

(1) Lateral surface: L.S. = $4e^2$ = 4 × e × e
(2) Total surface: T.S. = $6e^2$ = 6 × e × e

Sample Problems

(1) Find the lateral surface of a cube if its edge is 8 ft. long.

$$L.S. = 4e^2 = 4 \times e \times e$$
$$= 4 \times 8 \times 8$$
$$= 256$$

The lateral surface of the cube is 256 sq. ft.

(2) What is the total surface of a cube if its edge is 10 ft. long?

$$T.S. = 6e^2 = 6 \times e \times e$$
$$= 6 \times 10 \times 10$$
$$= 600$$

The total surface of the cube is 600 sq. ft.

Surfaces of Rectangular Solids

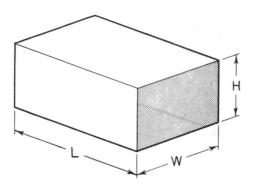

Information

 The lateral surface of a rectangular solid is equal to the sum of the areas of the front, back, and both ends.

 The total surface of a rectangular solid is equal to the sum of the areas of the front, back, top, bottom, and both ends.

Formulas

 (1) L.S. = LH + LH + WH + WH = 2LH + 2WH = 2H(L + W)

 (2) T.S. = LH + LH + WH + WH + LW + LW = 2LH + 2WH + 2LW

    = 2(LH + WH + LW)

Sample Problems

 (1) What is the lateral surface of a rectangular block 15 in. long, 9 in. wide, and 3 in. high.

   L.S. = 2H(L + W) = 2 × H × (L + W)

     = 2 × 3 × (15 + 9)

     = 6 × 24

     = 144

The lateral surface of the block is 144 sq. in.

 (2) Find the total surface of a rectangular block 20 in. long, 12 in. wide, and 7 in. high.

   T.S. = 2(LH + WH + LW)

     = 2 { (20 × 7) + (12 × 7) + (20 × 12) }

$$= 2(140 + 84 + 240)$$
$$= 2 \times 464$$
$$= 928$$

The total surface of the block is 928 sq. in.

Problems	Work Here

(1) How many sq. ft. of lumber will be required to construct the form for a cubical shaped footing 3 ft. on each edge?

Answer:

(2) Find the lateral surface of a ply-wood box with edges 18 in. long.

Answer:

(3) What is the total surface of a cubical bin whose edges are 9.5 ft. long?

Answer:

(4) How many square feet are there in the lateral surface of a chest 32 in. long, 24 in. wide, and 16 in. deep?

Answer:

(5) The form for a concrete footing is 10 ft. long, 22 in. wide, and 10 in. deep. How many square feet of lumber are there in this form?

Answer:

Problems	Work Here

(6) Determine the number of rolls of paper for the walls of a room 30 ft. long, 18 ft. wide, and 10 ft. high. Divide the wall area by 36 to get the number of rolls.

Answer:

(7) How many square feet of cedar will be required to line the walls, ceiling and floor of a closet $5\frac{1}{2}$ ft. long, 4 ft. wide, and 8 ft. high? Allow 20 per cent for waste.

Answer:

(8) How much wall primer must be used for the walls and ceilings of 10 rooms 24 ft. long, 15 ft. wide, and $9\frac{1}{2}$ ft. high if one gallon covers 800 sq. ft.?

Answer:

(9) Find the number of square feet of lumber required for the construction of a bin $22\frac{3}{4}$ ft. long, $18\frac{1}{2}$ ft. wide, and 10 ft. deep. Include a top and bottom for the bin.

Answer:

Surfaces of Cylinders

Study pages 429 to 431 of your textbook, Practical Mathematics, before preparing this test.

DO NOT SUBMIT THIS SELF-CHECK TEST FOR GRADING. Check your answers with the Answer Key at the end of this workbook.

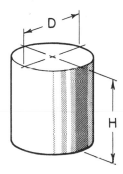

Information

The lateral surface of a cylinder is equal to the circumference multiplied by the height.

The total surface of a cylinder is equal to the lateral surface plus twice the area of the base.

Formulas

(1) L.S. = 3.1416DH = 3.1416 × D × H

(2) T.S. = 3.1416DH + 1.5708D^2 = (3.1416 × D × H) + (1.5708 × D × D)

Sample Problems

What is the lateral surface and total surface of a cylinder whose diameter is 2 ft. and length 14 ft.?

(1) L.S. = 3.1416 × 2 × 14
      = 87.9648

The lateral surface of the cylinder is 87.96 sq. ft. or 88 sq. ft.

(2) T.S. = 87.96 + 1.5708 × 2 × 2

       = 87.9648 + 6.2832

       = 94.248

The total surface of the cylinder is 94.25 sq. ft. or 94 1/4 sq. ft.

Problems	Work Here

(1) Find the lateral surface of a
cylindrical pipe whose diameter
is 15 in. and length 16 ft.
Hint: Change 15 in. to 1.25 ft.

Answer:

(2) What is the lateral surface of a
cylindrical silo whose diameter
is 14 ft. and height 30 ft.?

Answer:

(3) A cylindrical water tank has a
diameter of 15 ft. and a height
of 25.5 ft. What is its lateral
surface?

Answer:

(4) Determine the total surface of a
cylindrical bin whose diameter
is $5\frac{3}{4}$ ft. and height 15 ft.

Answer:

Surfaces of Pyramids

Study pages 433 to 435 of your textbook, <u>Practical</u> <u>Mathematics</u>, before preparing this test.

DO NOT SUBMIT THIS SELF-CHECK TEST FOR GRADING. Check your answers with the Answer Key at the end of this workbook.

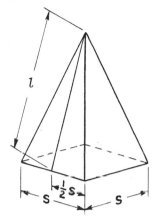

Information

The altitude of one of the triangles forming the faces of a pyramid is called the <u>slant height.</u>

The <u>lateral surface</u> of a pyramid is equal to the perimeter of its base multiplied by one-half the slant height.

The total surface of a pyramid is equal to the lateral surface plus the area of its base.

Formulas

(1) Area of face $= \dfrac{ls}{2}$

(2) Lateral surface: L.S. $= \dfrac{1}{2} lp = \dfrac{1}{2} \times l \times p$ (perimeter of base)

(3) Total surface: T.S. $= \dfrac{1}{2} lp + B = (\dfrac{1}{2} \times l \times p) + B$ (area of base)

Sample Problems

Find the lateral and total surface of a pyramid if the base is a square 8 ft. on each side and the slant height is 10 ft.

$$(1)\ \text{L.S.} = \dfrac{1}{\cancel{2}_1} \times \cancel{10}^{\,5} \times 32$$

$$= 160$$

The lateral surface is 160 sq. ft.

$$(2)\ \text{T.S.} = (\dfrac{1}{\cancel{2}_1} \times \cancel{10}^{\,5} \times 32) + (8 \times 8)$$

$$= 160 + 64$$

$$= 224$$

The total surface is 224 sq. ft.

Problems	Work Here

(1) Find the lateral surface of a roof in the form of a pyramid with a square base 12.5 ft. on each side if the slant height is 18 ft.

Answer:

(2) A wooden hopper in the form of a pyramid has a square base 6 ft. on each side. Find its lateral surface if the slant height is $9\frac{3}{4}$ ft.

Answer:

(3) How many square feet of copper are required to cover the lateral surface of a pyramidal tower whose base is a hexagon 6 ft. on each side if the slant height is 18 ft. ?

Answer:

(4) A concrete pyramid with a slant height of 37.5 ft. has a triangular base 10 ft. on each side. Find its total surface.

Answer:

Surfaces of Cones

Study pages 439 to 441 of your textbook, Practical Mathematics, before preparing this test.

DO NOT SUBMIT THIS SELF-CHECK TEST FOR GRADING. Check your answers with the Answer Key at the end of this workbook.

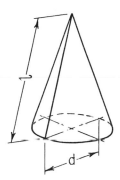

Information

The lateral surface of a cone is equal to the circumference of the base times one-half the slant height.

The total surface of a cone is equal to the lateral surface plus the area of the base.

Formulas

(1) L.S. $= \dfrac{\pi d\, l}{2} = \dfrac{3.1416 \times d \times l}{2} = 1.5708 \times d \times l$

(2) T.S. $= \dfrac{\pi d\, l}{2} + \dfrac{\pi d^2}{4} = (1.5708 \times d \times l) + (.7854 \times d \times d)$

Sample Problems

What is the lateral surface and total surface of a cone whose diameter is 5 ft. and slant height 12 ft.?

(1) L.S. $= 1.5708 \times 5 \times 12$
$= 94.248$

The lateral surface of the cone is 94 1/4 sq. ft.

(2) T.S. $= (1.5708 \times 5 \times 12) + (.7854 \times 5 \times 5)$
$= 94.248 + 19.635$
$= 113.883$

The total surface of the cone is $113\frac{7}{8}$ sq. ft. or approximately 114 sq. ft.

Problems	Work Here

(1) Determine the amount of sheet copper necessary to cover a conical shaped tower excluding the base if the diameter of the base is 20 ft. and the slant height is 20 ft.

Answer:

(2) Find the area of the roof of a cylindrical silo whose diameter is 20 ft. if the slant height is $11\frac{1}{4}$ ft.

Answer:

(3) How many square yards of painting must be done for one coat on a conical roof whose diameter is 30 ft. and slant height is 18 ft.?

Answer:

Surfaces of Frustums of Pyramids and Cones

Study pages 441 to 444 of your textbook, <u>Practical Mathematics</u>, before preparing this test.

DO NOT SUBMIT THIS SELF-CHECK TEST FOR GRADING. Check your answers with the Answer Key at the end of this workbook.

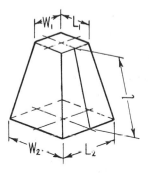

Frustum of a Pyramid

Information

      The lateral surface of the frustum of a pyramid is equal to one-half the sum of the perimeters of the two bases times the slant height.

      The total surface of the frustum of a pyramid is equal to the lateral surface plus the areas of the two bases.

Formulas

    (1) L.S. $= \dfrac{l\,(P_1 + P_2)}{2} = \dfrac{l \times (P_1 + P_2)}{2}$

    (2) T.S. $= \dfrac{l \times (P_1 + P_2)}{2} + B_1 + B_2$

Sample Problems

    (1) Find the lateral surface of the frustum of a pyramid whose bases are rectangular 10 in. by 14 in. and 15 in. by 21 in. if the slant height is 20 in.

$$\text{L.S.} = \frac{20 \times \left\{ (10 + 14 + 10 + 14) + (15 + 21 + 15 + 21) \right\}}{2}$$

$$= \frac{\overset{10}{\cancel{20}} \times (48 + 72)}{\underset{1}{\cancel{2}}}$$

$$= 10 \times 120$$

$$= 1200$$

The lateral surface of the frustum of the pyramid is 1,200 sq. in.

(2) Determine the total surface of a frustum of a pyramid whose bases are squares 15 in. and 25 in. on each side and whose slant height is 22 in.

$$\text{T.S.} = \frac{\overset{11}{\cancel{22}} \times (60 + 100)}{\underset{1}{\cancel{2}}} + 225 + 625$$

$$= (11 \times 160) + 225 + 625$$

$$= 1760 + 225 + 625$$

$$= 2610$$

The total surface of the frustum of the pyramid is 2,610 sq. in.

Problems	Work Here

(1) How many square feet of sheet metal are necessary to cover the sides of a bin whose bases are squares 5.5 ft. and 4.25 ft. on each side if the slant height is 6 ft.?

Answer:

(2) How many square feet of surface are found in an enclosed hopper if its bases are squares 10 ft. and 8 ft. on each side, and the slant height is 12 ft.?

Answer:

(3) The bases of a concrete support are regular hexagons 12 in. and 16 in. on each side. Find the total surface if their slant height is 20 in.

Answer:

Frustum of a Cone

Information

The lateral surface of the frustum of a cone is equal to one-half the circumference of the two bases multiplied by the slant height.

The total surface of the frustum of a cone is equal to the lateral surface plus the areas of the two bases.

Formulas

(1) L.S. $= \dfrac{l\,(C_1 + C_2)}{2} = \dfrac{3.1416 \times l \times (d_1 + d_2)}{2} = 1.5708 \times l \times (d_1 + d_2)$

(2) T.S. $= \dfrac{l \times (C_1 + C_2)}{2} + .7854\,(d_1{}^2 + d_2{}^2)$

$\qquad = 1.5708 \times l \times (d_1 + d_2) + .7854 \times (d_1{}^2 + d_2{}^2)$

Sample Problems

(1) What is the lateral surface of the frustum of a cone whose bases have diameters of 14 in. and 21 in. and slant height of 20 in.?

$\qquad$ L.S. $= 1.5708 \times 20 \times (14 + 21)$

$\qquad\qquad = 1.5708 \times 20 \times 35$

$\qquad\qquad = 1,099.56$

The lateral surface of the frustum of the cone is approximately 1,100 sq. in.

(2) What is the total surface of the frustum of a cone whose bases have diameters of 14 in. and 20 in. if the slant height is 22 in.

$\qquad$ T.S. $= 1.5708 \times 22 \times (14 + 20) + .7854 \times (196 + 400)$

$\qquad\qquad - 1.5708 \times 22 \times 34 + .7854 \times 596$

$\qquad\qquad = 1174.9584 + 468.0984$

$\qquad\qquad = 1643.0568$

The total surface of the frustum of the cone is approximately 1,643 sq. in.

(1) Determine the lateral surface of a
     wooden tub if the diameters of the
     top and bottom are 28 in. and 25.5
     in. and the slant height is 16 in.

     Answer:

(2) Find the lateral surface of a concrete
     support whose bases are circles with
     diameters of 18.75 in. and 23.5 in. if
     the slant height is 27 in.

     Answer:

(3) An open-top wooden vat is in the shape
     of the frustum of a cone. Find its sur-
     face if the diameter of the top base is
     20 ft. , the bottom base is 22 ft. , and
     the slant height is 6.5 ft.

     Answer:

(4) An open-top wooden vat is in the shape
     of the frustum of a cone. Find its sur-
     face if the diameters of the bases are
     22 ft. and 20 ft., and the slant height is
     6.5 ft.

     Answer:

Surfaces of Spheres and Hemispheres

Study pages 444 to 446 of your textbook, <u>Practical</u> <u>Mathematics</u>, before preparing this test.

DO NOT SUBMIT THIS SELF-CHECK TEST FOR GRADING. Check your answers with the Answer Key at the end of this workbook.

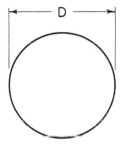 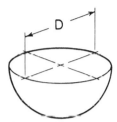

Information

  The surface of a sphere is equal to four times the area of a great circle or 3.1416 times the square of the diameter.

  The surface of a hemisphere is equal to two times the area of a great circle or 1.5708 times the square of the diameter.

Formulas

$$(1) \text{ L.S.} = 4\pi r^2 \text{ or } 4 \times 3.1416 \times r \times r$$
$$\text{L.S.} = 3.1416 d^2 \text{ or } 3.1416 \times d \times d$$

$$(2) \text{ L.S.} = 2\pi r^2 \text{ or } 2 \times 3.1416 \times r \times r$$
$$\text{L.S.} = 1.5708 d^2 \text{ or } 1.5708 \times d \times d$$

Sample Problems

  (1) Find the surface of a sphere whose diameter is 5 ft.

$$\text{L.S.} = 3.1416 \times 5 \times 5$$
$$= 78.54$$

The surface of the sphere is approximately 78.5 sq. ft.

  (2) What is the surface of a hemisphere whose diameter is 8 ft.?

$$\text{L.S.} = 1.5708 \times 8 \times 8$$
$$= 100.5312$$

The surface of the hemisphere is approximately 100.5 sq. ft.

Problems	Work Here

(1) Find the surface of a wooden sphere whose diameter is 12 in.

Answer:

(2) How many square feet are there in the surface of a spherical tank whose diameter is 20 ft.?

Answer:

(3) Determine the lateral surface of a hemispherical block whose diameter is 12.5 in.

Answer:

(4) How many square feet are there in the surface of a hemispherical dome whose diameter is 50 ft.?

Answer:

Examination No. 6

Based on pages 111 to 126 in this workbook

Student's Name_____Student Number_____

Street_____City_____State_____
Zip
Code_____

Problems	Show Work Here

(1) Find the total surface of a cubical box with top if the edges are 22 in. long.

Answer:

(2) Determine the lateral surface of the form for a cubical footing if the edges are 2.5 ft. long.

Answer:

(3) A footing is 25 ft. long, 25 in. wide and 11 in. deep. How many square feet are there in its surface?

Answer:

(4) How many square feet of surface are there in a bin 20 ft. long, 14 ft. wide, and $6\frac{1}{2}$ ft. deep? Include a top and bottom for the bin.

Answer:

(5) Find the total surface of a cylindrical supporting column whose length is $9\frac{1}{2}$ ft. and diameter 8 in.

Answer:

Problems	Show Work Here

(6) What is the lateral surface of a
wooden water tank 15 ft. deep if
its diameter is 10 ft.?

Answer:

(7) How many square feet are there in
the surface of a pyramid-shaped
roof whose base is a square 18 ft.
on each side if the slant height is
27 ft. ?

Answer:

(8) Find the lateral surface of a conical
tower whose slant height is 17 ft. if
the diameter of the base is 12 ft.

Answer:

(9) Determine the lateral surface of
a footing in the shape of the frustum
of a pyramid if the bases are squares
3 ft. and 4 ft. on each side, and the
slant height is 3 ft.

Answer:

(10) A hemispherical dome has a diameter
of 75 ft. How many square feet are
there in its surface?

Answer:

Volumes of Cubes and Rectangular Solids

Study pages 425 to 428 of your textbook, <u>Practical Mathematics</u>, before preparing this test.

DO NOT SUBMIT THIS SELF-CHECK TEST FOR GRADING. Check your answers with the Answer Key at the end of this workbook.

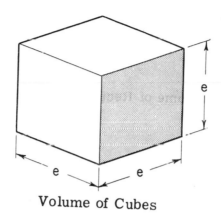

Volume of Cubes

Information

A <u>cube</u> is a prism whose bases and faces are all equal squares.

An <u>edge</u> is the line where two faces meet (e = edge).

The volume of a cube is equal to the cube of an edge.

Formula

$$V = e^3 \quad \text{or} \quad V = e \times e \times e$$

Sample Problems

(1) What is the volume of a cubical bin whose edges are 8 ft. long?

$$V = 8 \times 8 \times 8$$
$$= 512$$

The volume of the bin is 512 cu. ft.

(2) Determine the volume of a cubical box if the edges are 6 in long.

$$V = 6 \times 6 \times 6$$
$$= 216$$

The volume of the box is 216 cu. in.

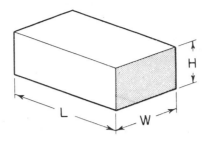

Volume of Rectangular Solids

Information

A <u>rectangular prism</u> is one whose bases and faces are all rectangles.

The volume of a rectangular prism is equal to the product of the length, width, and height.

Formula

$$V = LWH = L \times W \times H$$

Sample Problems

(1) What is the volume of a rectangular piece of wood 16 in. long, 8 in. wide, and 4 in. thick?

$$V = 16 \times 8 \times 4$$
$$= 512$$

The volume of the piece of wood is 512 cu. in.

(2) How much dirt must be removed from an excavation 50 ft. by 24 ft. by 6 ft.?

$$V = 50 \times 24 \times 6$$
$$= 7200$$

The amount of dirt to be removed is 7,200 cu. ft.

Problems	Work Here

(1) Find the volume of a cubical block of wood if its edges are 11 in. long.

Answer:

(2) What is the capacity of a cubical bin whose edges are 5.5 ft. long?

Answer:

(3) A cubical vat whose edges are 10 ft. long is filled with water. Find the number of gallons in the vat if there are $7\frac{1}{2}$ gal. in 1 cu. ft.

Answer:

(4) Find the weight of a cube of maple wood with 8-inch edges if maple weighs 48 lb. per cu. ft.

Answer:

(5) How many cubic feet of pine are there in a piece of lumber 15 in. wide, 11 in. thick, and 20 ft. long?

Answer:

Problems	Work Here

(6) Find the number of cubic feet of air in a room 125 ft. long, 80 ft. wide and 20.5 ft. high.

Answer:

(7) Determine the number of cubic yards of earth to be removed for a cellar 45 ft. long, 24 ft. wide, and 8 ft. deep.

Answer:

(8) How many gallons of water are there in a tank 75 ft. long and 38 ft. wide if it is filled to a depth of 5 ft.? There are $7\frac{1}{2}$ gal. in 1 cu. ft.

Answer:

(9)

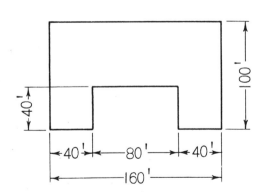

How many cubic yards of earth must be removed for this excavation if the depth is 8 ft.?

Answer:

Volume of Cylinders

Study pages 430 to 433 of your textbook, <u>Practical</u> <u>Mathematics</u>, before preparing this test.

DO NOT SUBMIT THIS SELF-CHECK TEST FOR GRADING. Check your answers with the Answer Key at the end of this workbook.

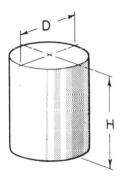

Information

A <u>right circular cylinder</u> is one whose lateral surface is perpendicular to the base. The bases are parallel and circular.

The volume of a cylinder is found by multiplying the area of its base by its altitude.

Formula

$$V = .7854D^2 \, H = .7854 \times D \times D \times H$$

Sample Problems

(1) What is the capacity of a cylindrical bin whose diameter is 22 ft. and depth 30 ft.?

$$V = .7854 \times 22 \times 22 \times 30$$
$$= 11404.008$$

The capacity of the bin is 11,404 cu. ft.

(2) Find the volume of a cylindrical support 14 ft. long if its diameter is 15 in.

$$V = .7854 \times 1.25 \times 1.25 \times 14$$
$$= 17.18$$

The volume of the support is 17.18 cu. ft. or 17.2 cu. ft.

Problems	Work Here

(1) How many cubic yards of concrete
are necessary to build a solid
cylindrical support 18 ft. high
with a diameter of 6 ft. ?

Answer:

(2) Determine the capacity in gallons
of a cylindrical wooden vat 20 ft.
in diameter and $4\frac{1}{2}$ ft. deep. A
cubic foot contains about $7\frac{1}{2}$ gal.

Answer:

(3) Find the number of cubic yards of
earth to be excavated for a cistern
15 ft. deep with a diameter of 10 ft.

Answer:

(4) How many cubic feet of material are
there in a cylindrical timber 22 ft.
long if its diameter is 15 in. ?

Answer:

Volumes of Pyramids

Study pages 435 to 438 of your textbook, <u>Practical</u> <u>Mathematics</u>, before preparing this test.

DO NOT SUBMIT THIS SELF-CHECK TEST FOR GRADING. Check your answers with the Answer Key at the end of this workbook.

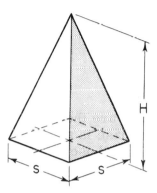

Information

      The altitude of a pyramid is the perpendicular distance from the vertex to the base.

      The volume of a pyramid is equal to the product of one-third its base area and its altitude.

Formula

$$V = \frac{1}{3}\,BH \;=\; \frac{B \times H}{3}$$

Sample Problems

      (1) Find the volume of a pyramid whose base is a square 2.5 ft. on each side and whose height is 6 ft.

$$V = \frac{2.5 \times 2.5 \times \overset{2}{\cancel{6}}}{\underset{1}{\cancel{3}}}$$

$$= 12.5$$

The volume of the pyramid is 12.5 cu. ft.

      (2) What is the volume of a pyramidal block whose base is a rectangle 14 in. long and 9 in. wide if its height is 17.5 in.?

$$V = \frac{14 \times \overset{3}{\cancel{9}} \times 17.5}{\underset{1}{\cancel{3}}}$$

The volume of the block is 735 cu. in.

Problems	Work Here

(1) Find the volume of a pyramidal block whose base is a square 15 in. on each side and whose height is 22 in.

Answer:

(2) How many cubic feet of space are there in a pyramidal tower whose base is a square 18 ft. on each side and whose height is 37.5 ft.?

Answer:

(3) Determine the number of cubic yards of concrete in a pyramid whose base is a square 8 yd. on each side and whose altitude is 42 ft.

Answer:

(4) Find the capacity in gallons of a pyramidal tank whose base is a square 12.5 ft. on each side and whose depth is 22.5 ft.

A cubic foot contains about $7\frac{1}{2}$ gal.

Answer:

Volumes of Cones

Study pages 439 to 441 of your textbook, <u>Practical</u> <u>Mathematics</u>, before preparing this test.

DO NOT SUBMIT THIS SELF-CHECK TEST FOR GRADING. Check your answers with the Answer Key at the end of this workbook.

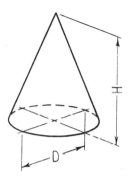

Information

The altitude of a cone is the perpendicular distance from the vertex to the base.

The volume of a cone is the product of one-third its base area and its altitude.

Formula

$$V = \frac{1}{3}BH = \frac{B \times H}{3}$$

$$V = \frac{.7854D^2 \times H}{3} = .2618 \times D \times D \times H$$

Sample Problems

(1) Determine the volume of a cone whose diameter is 2 ft. and height 3.5 ft.

$$V = .2618 \times 2 \times 2 \times 3.5$$

$$= 3.6652 \text{ cu. ft.}$$

The volume of the cone is approximately **3.67 cu. ft. or 3.7 cu. ft.**

(2) Find the number of cubic yards of sand in a conical pile if the diameter is 25 ft. and the height 18 ft.

$$V = \frac{\overset{.2618}{\cancel{.7854}} \times 25 \times 25 \times \overset{2}{\cancel{18}}}{\underset{1}{\cancel{3}} \times \underset{3}{\cancel{27}}}$$

$$= 109.08$$

The pile of sand contains **109 cu. yd.**

Problems	Work Here

(1) Determine the number of cubic
yards of sand in a pile which has
the shape of a cone if the diameter
of its base is 25 ft. and its height
10 ft.

Answer:

(2) What is the capacity in gallons of a
conical tank 15 ft. in diameter and
10.5 ft. deep? There are approximately $7\frac{1}{2}$ gal. per cu. ft.

Answer:

(3) A concrete cone is to be placed on top
of a cylindrical column whose diameter
is the same as the cone, 3 ft. 3 in.
How many cubic yards of concrete will
be required for the cone if it is 12 ft.
high?

Answer:

Volumes of Frustums of Pyramids and Cones

Study pages 441 to 444 of your textbook, <u>Practical Mathematics</u>, before preparing this test.

DO NOT SUBMIT THIS SELF-CHECK TEST FOR GRADING. Check your answers with the Answer Key at the end of this workbook.

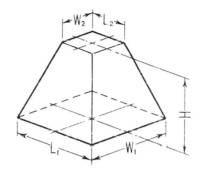

Volumes of Frustums of Pyramids

Information

    If the top of a pyramid is cut off by a plane parallel to the base, the part remaining is called a <u>frustum.</u>

    The volume of the frustum of a pyramid is equal to the sum of the areas of the two bases plus the square root of the product of the areas of the two bases with this sum multiplied by one-third of the altitude.

Formula

$$V = \frac{H}{3}(B_1 + B_2 + \sqrt{B_1 B_2}) \quad = \quad \frac{H \times (B_1 + B_2 + \sqrt{B_1 \times B_2})}{3}$$

Sample Problem

    What is the volume of a concrete supporting block in the shape of a frustum of a pyramid with square bases 21 in. and 28 in. on each side if the height is 27 in.?

$$V = \frac{\overset{9}{\cancel{27}} \times (441 + 784 + \sqrt{441 \times 784})}{\underset{1}{\cancel{3}}}$$

$$= 9 \times (441 + 784 + \sqrt{345744})$$

$$= 9 \times (441 + 784 + 588)$$

$$= 9 \times 1813$$

$$= 16317$$

The volume of the frustum is 16,317 cu. in.

Problems	Work Here

(1) Determine the amount of concrete
in cubic feet necessary for 10 concrete pyramidal pieces if the bases
are squares 15 in. and 20 in. on
each side, and the height is 27 in.

Answer:

(2) Find the capacity of a pyramidal
bin whose upper edges are 8 ft.
long and lower edges 6 ft. long if
its depth is 10 ft.

Answer:

(3) The upper and lower bases of a
group of concrete pyramidal pieces
are regular hexagons each 8 in. and
10 in. on a side with an altitude of
15 in.   How many cubic feet of concrete will be required for 80 of
these pieces ?

Answer:

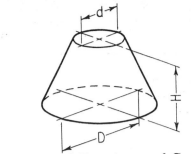

Volumes of Frustums of Cones

Information

If the top of a cone is cut off by a plane parallel to the base, the part remaining is called a frustum.

The volume of the frustum of a cone is equal to the sum of the squares of
the diameters of the bases, plus the product of the diameters of the bases with
this sum multiplied by .2618 times the altitude.

Formula

$$V = .2618H\ (D^2 + d^2 + Dd) = .2618 \times H \times (D^2 + d^2 + Dd)$$

Sample Problem

The diameters of the bases of the frustum of a cone are 18 in. and 32 in.
and its height 26 in. What is its volume?

$$V = .2618 \times 26 \times (1024 + 324 + 576)$$

$$= .2618 \times 26 \times 1924$$

$$= 13096.2832$$

The volume of the frustum is 13,096.28 cu. in. or approximately

$13,096\frac{1}{3}$ cu. in.

Problems	Work Here

(1) Find the capacity of a wooden tub
if the diameters of the bases are
21 in. and 18 in., and the depth
16 in.

Answer:

(2) What is the capacity of a bin shaped
like the frustum of a cone if the
diameters of the upper and lower
bases are 12 ft. and 10 ft., and the
depth is 12 ft.?

Answer:

(3) A contract calls for 100 concrete
pieces having the shape of frustums
of right circular cones. Find the
amount of material required if the
diameters of the bases are 20 in.
and 30 in., and the height 38 in.

Answer:

Volumes of Spheres and Hemispheres

Study pages 446 to 448 of your textbook, <u>Practical</u> <u>Mathematics</u>, before preparing this test.

DO NOT SUBMIT THIS SELF-CHECK TEST FOR GRADING. Check your answers with
the Answer Key at the end of this workbook.

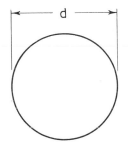 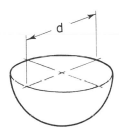

Information

The volume of a sphere is equal to the cube of the diameter multiplied by
.5236.

The volume of a hemisphere is equal to the cube of the diameter multiplied by .2618.

Formulas

(1) Sphere: $V = .5236D^3 = .5236 \times D \times D \times D$

(2) Hemisphere: $V = .2618 \ D^3 = .2618 \times D \times D \times D$

Sample Problems

(1) What is the volume of a wooden ball whose diameter is 8 in.?

$$V = .5236 \times 8 \times 8 \times 8$$

$$= 268.0832$$

The volume of the ball is approximately 268 cu. in.

(2) Find the volume of a hemispherical tank whose diameter is 25 ft.

$$V = .2618 \times 25 \times 25 \times 25$$

$$= 4,090.625$$

The volume of the hemispherical tank is 4,090.625 cu. ft.

Problems	Work Here

(1) Find the volume of a wooden ball whose diameter is 10 in.

Answer:

(2) What is the capacity in gallons of a spherical tank whose diameter is 22.5 ft.? A cubic foot contains about $7\frac{1}{2}$ gal.

Answer:

(3) Determine the volume of a hemispherical block whose diameter is 18 in.

Answer:

(4) Find the number of cubic yards of air space inside a hemispherical dome with a diameter of 125 ft.

Answer:

Examination No. 7

Based on pages 129 to 144 in this workbook

Student's Name_____Student Number_____

Street_____City_____State_____Zip Code_____

Problems	Show Work Here

(1) A cubical block of wood has edges 18 in. long. Find its volume.

Answer:

(2) Find the capacity in gallons of a cubical water tank whose edges are 16 ft. long. There are $7\frac{1}{2}$ gal. in 1 cu. ft.

Answer:

(3) How many cubic feet of walnut are there in a timber 20 ft. long, 18 in. wide, and 15 in. thick?

Answer:

(4) Find the number of cubic yards of air in a room 25 ft. long, 15 ft. wide, and $9\frac{1}{2}$ ft. high. There are 27 cu. ft. in 1 cu. yd.

Answer:

(5) Determine the number of cubic yards in a concrete column 10 ft. high if its diameter is 27 in.

Answer:

Problems   Show Work Here

(6) The base of a pyramidal block
    is a square with 16 in. sides.
    Find its volume if its height is
    22 in.

    Answer:

(7) A conical-shaped pile of sand has
    a diameter of 37 ft. at its base.
    How many cubic yards are there
    in the pile if it is 18 ft. high?

    Answer:

(8) Determine the volume of a bin
    shaped like the frustum of a
    pyramid whose bases are squares
    15 ft. and 12 ft. on each side. The
    bin is 8 ft. deep.

    Answer:

(9) Find the capacity of a wooden vat
    whose shape is that of the frustum
    of a cone. The diameters are 25 ft.
    and 23 ft., and the depth is 6 ft.

    Answer:

(10) What is the volume of a wooden
     sphere whose diameter is 6 in. ?

     Answer:

Lumber Measurement
Regular Shapes

DO NOT SUBMIT THIS SELF-CHECK TEST FOR GRADING. Check your answers with
the Answer Key at the end of this workbook.

Information

The unit of measurement for lumber is the <u>board foot</u>. This unit of
measure is equal to a board 1 ft. square and 1 in. thick.

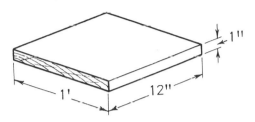

How To Figure Board Feet

The number of board feet in a piece of lumber is found by multiplying the
length in feet by the width in inches by the thickness in inches and then dividing the result by 12.

$$\text{Board Feet} = \frac{\text{length in feet} \times \text{width in inches} \times \text{thickness in inches}}{12}$$

$$\text{B.F.} = \frac{l \times w \times t}{12} \text{ where } l \text{ equals the length in feet, w equals the width} \\ \text{in inches, and t the thickness in inches.}$$

Sample Problems

(1) Find the number of board feet in a piece of lumber 18 ft. long, 6 in.
wide, and 1 in. thick.

Instructions	Computations
(a) Write the formula.	(a) $\text{B.F.} = \frac{l \times w \times t}{12}$
(b) Write the given values.	(b) $l = 18,\ w = 6,\ t = 1$
(c) Substitute these values in the formula.	(c) $\text{B.F.} = \frac{18 \times 6 \times 1}{12}$
(d) Cancel.	(d) $\text{B.F.} = \frac{\overset{3}{\cancel{18}} \times \overset{3}{\cancel{6}} \times 1}{\underset{\underset{1}{2}}{\cancel{12}}}$
(e) Multiply.	(e) $\text{B.F.} = 9$

There are 9 bd. ft. in the piece of lumber.

(2) Determine the number of board feet of lumber in 15 pieces 20 ft. long, 10 in. wide, and 2 in. thick.

Instructions	Computations
(a) Write the formula	(a) B.F. $= \dfrac{l \times w \times t}{12}$
(b) Write the given values	(b) $l = 20, \ w = 10, \ t = 2$
(c) Substitute these values in the formula.	(c) B.F. $= \dfrac{20 \times 10 \times 2}{12}$
(d) Cancel.	(d) B.F. $= \dfrac{\overset{5}{\cancel{20}} \times 10 \times 2}{\underset{3}{\cancel{12}}}$
(e) Multiply.	(e) B.F. $= \dfrac{100}{3}$
(f) Multiply by the number of pieces.	(f) B.F. $= \overset{5}{\cancel{15}} \times \dfrac{100}{\underset{1}{\cancel{3}}}$
(g) Multiply.	(g) B.F. = 500

There are 500 bd. ft. in the 15 pieces of lumber.

Problems	Work Here

Find the number of board feet of lumber in each of these problems.

(1) t = 1 in., w = 5 in., l = 24 ft.

Answer:

(2) t = 2 in., w = 8 in., l = 18 ft.

Answer:

(3) t = 3 in., w = 10 in., l = 30 ft.

Answer:

Problems	Work Here

(4) 6 pieces: t = 1 in., w = 4 in., l = 12 ft.

Answer:

(5) 20 pieces: t = 2 in., w = 6 in., l = 18 ft.

Answer:

(6) 42 pieces: t = 4 in., w = 8 in., l = 16 ft.

Answer:

(7) 50 pieces: t = 6 in., w = 10 in., l = 20 ft.

Answer:

(8) 32 pieces: t = 2 in., w = 4 in., l = 14 ft.

Answer:

SELF-CHECK TEST 52

Lumber Measurement
Irregular Shapes

DO NOT SUBMIT THIS SELF-CHECK TEST FOR GRADING. Check your answers with the Answer Key at the end of this workbook.

Information

There may be occasions when it is necessary to determine the number of board feet in pieces with irregular shapes such as the parallelogram, trapezoid, and triangle.

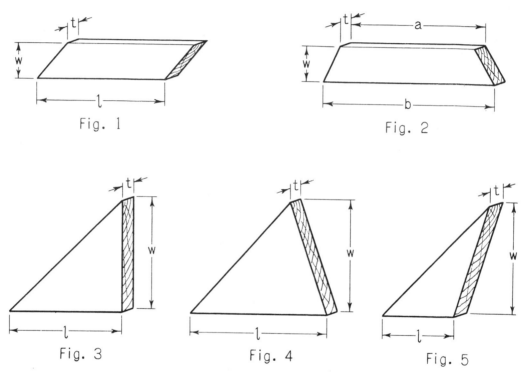

Fig. 1 Fig. 2 Fig. 3 Fig. 4 Fig. 5

Formulas

Fig. 1: B.F. = $\dfrac{l \times w \times t}{12}$ where l is the length in feet, w the width in inches, and t the thickness in inches.

Fig. 2: B.F. = $\dfrac{(a + b) \times w \times t}{24}$ where (a + b) is the sum of the lengths of the parallel sides, w the width in inches, and t the thickness in inches.

Figs. 3, 4 and 5: B.F. = $\dfrac{l \times w \times t}{24}$ where l is the length in feet, w the width in inches, and t the thickness in inches.

Sample Problems

(1) Find the number of board feet of material in a timber 18 ft. long, 14 in. wide, and 8 in. thick if the shape is that of Fig. 1.

$$B.F. = \frac{\overset{3}{\cancel{18}} \times \overset{7}{\cancel{14}} \times 8}{\underset{\underset{1}{\cancel{2}}}{\cancel{12}}}, \quad B.F. = 168$$

There are 168 bd. ft. of material in the timber.

(2) How many board feet are there in a piece 3 in. thick if the parallel sides are 10 ft. and 14 ft. long, and the width is 6 in.?

$$B.F. = \frac{(a + b) \times w \times t}{24}$$

$$B.F. = \frac{(10 + 14) \times 6 \times 3}{24}, \quad B.F. = \frac{\overset{1}{\cancel{24}} \times 6 \times 3}{\underset{1}{\cancel{24}}}, \quad B.F. = 18$$

There are 18 bd. ft. in the piece of material.

(3) A triangular shaped piece of lumber is 12 ft. long, its greatest width is 14 in., and its thickness 2 in. Find the number of board feet in this piece.

$$B.F. = \frac{l \times w \times t}{24}$$

$$B.F. = \frac{\overset{1}{\cancel{12}} \times 14 \times \overset{1}{\cancel{2}}}{\underset{\underset{1}{\cancel{2}}}{\cancel{24}}}, \quad B.F. = 14$$

There are 14 bd. ft. in this piece of lumber.

(4) Find the number of board feet in a piece 1 in. thick whose parallel sides are 15 in. and 21 in. long if the width of the piece is 15 in. Change 15 in. and 21 in. to feet.

$$\text{Side a: 15 in.} = 1\frac{1}{4} \text{ ft. or 1.25 ft.}$$

$$\text{Side b: 21 in.} = 1\frac{3}{4} \text{ ft. or 1.75 ft.}$$

$$B.F. = \frac{(a + b) \times w \times t}{24}$$

$$B.F. = \frac{(1.25 + 1.75) \times 15 \times 1}{24}, \quad B.F. = \frac{\overset{1}{\cancel{3}} \times 15 \times 1}{\underset{8}{\cancel{24}}}, \quad B.F. = \frac{15}{8} \text{ or } 1\frac{7}{8}$$

There are $1\frac{7}{8}$ bd. ft. in this piece of lumber.

Problems	Work Here

(1) Find the number of board feet of
material in 84 pieces cut in the
shape of Fig. 1, page 138, if the
length of each piece is 18 ft., the
width 4 in., and the thickness 2 in.

Answer:

(2) Thirty pieces of 1-inch material
8 in. wide with parallel sides 18 ft.
and 22 ft. long are required for a
certain job. Determine the board
feet of material.

Answer:

(3) Forty-five pieces of 2-inch lumber
12 in. wide and 2 ft. long with the
shape shown in Fig. 3, page 138,
are to be cut for a certain job. How
many board feet of material are required?

Answer:

(4) How many board feet of material will
be used for three dozen pieces cut in
the shape of Fig. 5, page 138, if each
piece is 27 in. long, 10 in. wide, and
3 in thick?

Answer:

Formwork

Information

Each class of formwork is figured separately. In general, formwork is
figured in square feet of contact area.

Foundation Footings

Foundation footings must have formwork on both sides (inside and outside). The lengths of the inside and outside forms are multiplied by the depth
of the footing to obtain the square feet of contact area. More simply, outside
perimeter times depth plus inside perimeter times depth equals square feet of
contact area. Since the outside footings extend 6 inches beyond the foundation,
1 foot must be added to the length and to the width.

Sample Problem 1

Find the square feet of contact area for the foundation footings in this
figure. Forms are indicated in dotted lines. The footings are 1 foot deep.

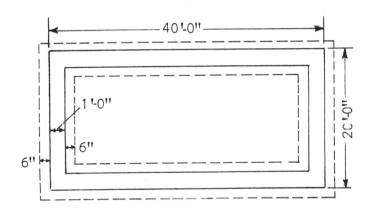

First find the square feet of contact area of the outside forms.
    Find the length of the forms. 40'-0" + 1'-0" = 41'-0"
    Find the width of the forms. 20'-0" + 1'-0" = 21'-0"
    Multiply the perimeter by the depth: 41'-0"
                                        21'-0"
                                        41'-0"
                                        <u>21'-0"</u>
                         124'-0" × 1'-0" = 124'-0"

Next find the square feet contact area of the inside forms.
    Find the length of the forms: 40'-0" - 1'-6" - 1'-6" = 37'-0"
    Find the width of the forms: 20'-0" - 1'-6" - 1'-6" = 17'-0"
    Multiply the perimeter by the depth: 108'-0" × 1'-0" = 108'-0"

Add the inside and outside forms: 124 sq. ft. and 108 sq. ft. = 232 sq. ft.

Column Footings

Formwork for interior column footing is figured in the same manner as the foundation footings. Since column footings are usually solid, only an outside form will be required.

Sample Problem 2

Find the amount of formwork required for a column footing 2'-0" square by 1'-0" deep.

Perimeter of the footing times the depth = 8'-0" × 1'-0" = 8 sq. ft.

Foundation Walls

Foundation walls are figured in the same manner as the foundation footings. Forms for the foundation wall do not extend as the footings, however.

Sample Problem 3

Using the figure in Sample Problem 2, find the square feet of contact area for the foundation walls, which are 7'-0" deep.

Find the area of the outside form.

40'-0" + 20'-0" + 40'-0" + 20'-0" × 7'-0" = 840 sq. ft.

Find the area of the inside forms.

38'-0" + 18'-0" + 38'-0" + 18'-0" × 7'-0" = 784 sq. ft.
Add the forms: 840 sq. ft. + 784 sq. ft. = 1624 sq. ft.

Problems	Work Here

(1-2) A foundation 13'-0" square, with walls 1 foot thick, requires footings 1 foot deep that extend 6" on each side of the foundation. What is the square feet of contact area required for the footing formwork? What is the square feet of contact area required for the foundation 6 feet deep?

Answer:

Answer:

(3) Find the square feet of contact area for two solid columns 3'-0" square by 8'-0".

Answer:

Sills

DO NOT SUBMIT THIS SELF-CHECK TEST FOR GRADING. Check your answers with the Answer Key at the end of this workbook.

Information

A <u>sill</u> is a timber placed horizontally on the foundation wall forming the lowest part of the frame which is to be supported. It is also the lowest part beneath any opening such as a door, window, etc.

The sill is usually laid 1 in. from the outer face of the foundation wall. It should be set into the concrete and securely fastened by bolts.

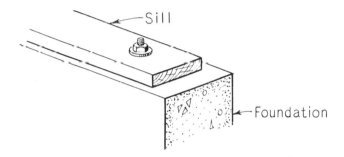

How To Determine the Amount of Material

Material for sills is sold in multiples of 2-foot lengths. It is advisable to avoid excessive lengths to eliminate the possibility of bending and warping.

In estimating the lineal feet of sills required for a job, disregard distance such as 1 in. from the outer face of the foundation wall, overlap at corners, and 1 ft. on each length for splicing. Determine the lineal feet of sills by finding the perimeter of the structure. The perimeters of irregularly shaped foundations may be found by adding the lengths of all walls.

Sample Problem 1

How many lineal feet of sill are required for a house 34 ft. long and 24 ft. wide?

Instructions	Computations
(1) Write the formula.	(1) $P = (2 \times l) + (2 \times w)$
(2) Write the given values.	(2) $l = 34$, $w = 24$
(3) Substitute these values in the formula.	(3) $P = (2 \times 34) + (2 \times 24)$ $= 68 + 48 = 116$

The lineal feet of sill required is 116.

Sample Problem 2

Find the number of 18-ft. pieces of sill required for a building 34 ft. long and 22 ft. wide.

Instructions	Computations
(1) Write the formula.	(1) $P = (2 \times l) + (2 \times w)$
(2) Write the given values.	(2) $l = 34$, $w = 22$
(3) Substitute these values in the formula. The lineal feet of sill is 112.	(3) $P = (2 \times 34) + (2 \times 22)$ $= 68 + 44$ $= 112$
(4) To find the number of pieces divide the linear feet by the length of each piece.	(4) $112 \div 18 = 6.2$

Order seven 18-ft. pieces of sill.

Problems	Work Here

---

(1) Find the lineal feet and number of sills required for this foundation. Use 22-ft. lengths.

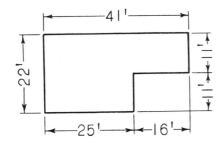

Answer:

(2) How many sills should be ordered for this foundation? Use 16-ft. lengths.

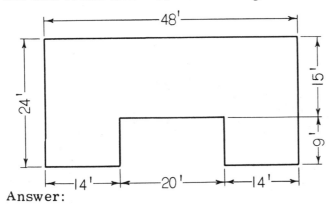

Answer:

Girders

DO NOT SUBMIT THIS SELF-CHECK TEST FOR GRADING. Check your answers with the Answer Key at the end of this workbook.

Information

A <u>girder</u> is a large built-up piece of lumber, steel, or other material used to support walls or joists over an opening. These horizontal members are usually supported at the ends by the foundation of the structure. Between the foundation walls, pieces or columns are used to provide support.

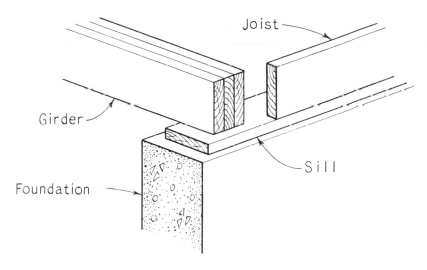

In making built-up girders, three or more 2-in. pieces placed as shown above are used. For example, in constructing an 8-in. by 10-in. girder, four pieces 2 in. by 10 in. are used.

Complete this table.

Size of Girder	Number of Two-Inch Pieces
6 in. by 8 in.	
10 in. by 12 in.	
12 in. by 14 in.	
10 in. by 14 in.	
8 in. by 10 in.	
8 in. by 12 in.	

Sample Problem

Find the number of board feet of material required for a built-up girder 8 in. thick, 10 in. wide, and 15 ft. long.

$$\text{B.F.} = \frac{\overset{2}{\cancel{8}} \times 10 \times \overset{5}{\cancel{15}}}{\underset{\underset{1}{\cancel{3}}}{\cancel{12}}}, \quad \text{B.F.} = 100$$

This girder requires 100 bd. ft. of material.

Problems	Work Here

(1) How many board feet of material are required for 5 built-up girders 6 in. thick, 10 in. wide, and 12 ft. long?

Answer:

(2) Calculate the number of board feet of material required for 16 built-up girders 12 in. thick, 14 in. wide, and 16 ft. long.

Answer:

(3) Determine the number of board feet of material in 10 supporting columns 4 in. thick, 6 in. wide, and 7 ft. 6 in. high.

Answer:

(4-6) Find the number of board feet of material for the following pieces:

Number of Pieces	Dimensions
(4) 10	16 ft. × 10 in. × 12 in.
(5) 18	8 ft. × 6 in. × 6 in.
(6) 4	24 ft. × 8 in. × 8 in.

Answer:

Answer:

Answer:

(7) Find the total of the above pieces.

Answer:

Floor Joists

DO NOT SUBMIT THIS SELF-CHECK TEST FOR GRADING. Check your answers with the Answer Key at the end of this workbook.

Information

Joist lengths for the first floor are estimated by examining the basement plan. Joists are usually placed 16 in. center to center, although other spacings, such as 12 in. and 24 in., are used depending upon the strength desired.

Allowance must be made for double framing around openings such as the fireplace hearth, stair wells, etc. To support the added weight of bathroom fixtures, the joists underneath must be doubled or the amount of space between them decreased. Joists are doubled under first-floor and second-floor partitions.

After finding the number of joists required, suitable lengths should be selected so as to avoid excessive waste.

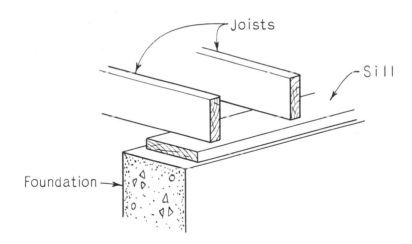

How To Determine the Number of Joists

The number of joists required for any given length may be determined by using

    (1) a fraction found by dividing 12 by the required center to center spacing,
    (2) then multiplying the given length by this fraction,
    (3) and adding one to this number.

Sample Problem

How many joists are required for a building 40 ft. long if they are placed 16 in. center to center?

Instructions	Computations
(1) Divide 12 by 16.	(1) $\dfrac{\overset{3}{\cancel{12}}}{\underset{4}{\cancel{16}}} = \dfrac{3}{4}$
(2) Multiply 40 by $\dfrac{3}{4}$	(2) $\overset{10}{\cancel{40}} \times \dfrac{3}{\underset{1}{\cancel{4}}} = 30$
(3) Add 1 to 30	(3) $30 + 1 = 31$

The number of pieces required is 31.

Problems	Work Here

(1) Determine the number of joists spaced 20 in. center to center for a building 50 ft. long.

Answer:

(2) Find the number of joists spaced 18 in. center to center for a building 42 ft. long.

Answer:

(3) How many joists spaced 24 in. center to center will be required for a building 54 ft. long?

Answer:

Examination No. 8

Based on pages 147 to 160 in this workbook

Student's Name_____ Student Number_____

Street_____City_____State_____
Zip
Code_____

| | |
| Problems | Show Work Here |

(1) How many board feet are there in 10 joists 2 in. by 10 in. if the length of each is 16 ft.?

Answer:

(2) A certain job requires 50 pieces 2 in. by 4 in. whose lengths are $9\frac{1}{2}$ ft. Find the number of board feet of 2 by 4's.

Answer:

(3) Twelve girders each 10 ft. long are built up of 3 pieces 2 in. by 8 in. Determine the number of board feet in the girders.

Answer:

(4) Fifteen shelves 1 in. thick and 10 in. wide with parallel sides $9\frac{1}{2}$ ft. and 10 ft. long are required for a job. How many board feet are there in the shelves?

Answer:

(5) How many board feet of material are there in 47 pieces of lumber 15 ft. long in the shape of parallelograms if each piece is 8 in. wide and 2 in. thick?

Answer:

Problems	Show Work Here

(6) Determine the number of board feet of lumber in 25 triangular pieces 11 ft. long if the greatest width of each piece is 10 in. and the thickness is $1\frac{1}{2}$ in.

Answer:

(7) How many lineal feet of sill are required for a building 40 ft. long and 26 ft. wide?

Answer:

(8) If 16-foot pieces of sill are used, find the number of pieces for a building 30 ft. long and 22 ft. wide.

Answer:

(9) Find the square feet of contact area for a solid column $2\frac{1}{2}$ ft. by 4 ft. by 12 ft. high.

Answer:

(10) Find the square feet of contact area required for a foundation whose inside dimensions are 30 ft. long, 24 ft. wide and 4 ft. deep. The walls are 1 ft. thick.

Answer:

Student's Name_____Student Number_____

Problems	Show Work Here

(11) A certain job required 4 built-up
girders 18 ft. long. Each is made
of 3 pieces 10 in. wide and 2 in.
thick. Find the number of board
feet of material required.

Answer:

(12-15) A contractor has to order sills
for three different jobs. One building is 52 ft. long and 32 ft. wide,
another is 44 ft. long and 30 ft.
wide, and the third is 38 ft. long
and 26 ft. wide. Find the lineal
feet of sills for each building and
the total for all three.

Answer:

Answer:

Answer:

Answer Total:

(16) How many joists spaced 18 in. center
to center are required for a building
39 ft. long?

Answer:

(17) The joists for a building are to be
spaced 16 in. center to center. How
many are needed for a building 36 ft.
long?

Answer:

Problems	Show Work Here

(18-19) How many lineal feet of sills
and how many joists are required
for this foundation? The joists are
spaced 16 in. on center.

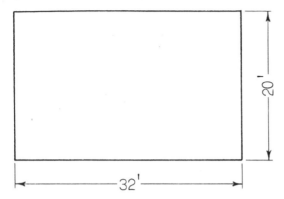

Answer:

Answer:

(20) How many sills 18 ft. long should be
ordered for this foundation?

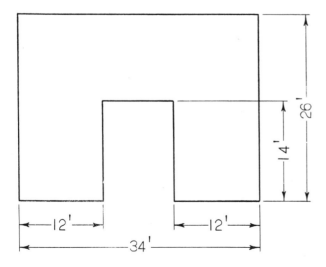

Answer:

Bridging

Information

Bridging is used to strengthen framing. It is placed between joists and
partition studs to make them more rigid and to hold them in place.

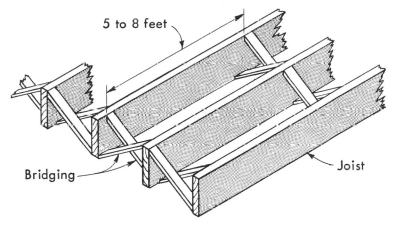

5 to 8 feet

Bridging

Joist

Two pieces of bridging are required for each joist. Rows of bridging
are usually spaced 5 to 8 feet apart. Since bridging is sold by pieces or by
lineal length, no board feet measure is necessary.

The number of spaces between joists is one less than the number of joists.
To find the number of pieces of bridging, multiply the number of spaces by 2.
For example, if the number of spaces is 18, then the number of pieces is
18 × 2 = 36 pieces.

How To Find the Length of a Piece of Bridging

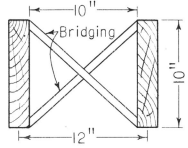

Bridging

For 2 by 10 joists on 12-in. centers, the distance between the joists is
10 in. Although the 2 by 10's are actually $1\frac{1}{2}$ by $9\frac{1}{4}$, we will consider the
width as 10 in.

The width of the joist, the distance between joists, and the length of the piece of bridging can be considered as the sides of a right triangle.

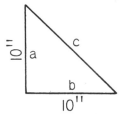

For any right triangle, $c^2 = a^2 + b^2$ or $c = \sqrt{a^2 + b^2}$, this formula can be used in determining the lengths of pieces of bridging.

Sample Problems

(1) Find the length of a piece of bridging for 2 by 10 joists on 12 in. centers.

Instructions	Computations
(a) Write the formula.	(a) $c = \sqrt{a^2 + b^2}$
(b) Write the formula in this form.	(b) $c = \sqrt{(a \times a) + (b \times b)}$
(c) Write the given value.	(c) $a = 10, \quad b = 10$
(d) Substitute these values in the formula.	(d) $c = \sqrt{(10 \times 10) + (10 \times 10)}$
(e) Multiply, then add.	(e) $c = \sqrt{100 + 100}$ $= \sqrt{200}$
(f) Find the square root.	(f) $c = 14.142$

Each piece of bridging is approximately 14 in. long.

(2) Approximately how many lineal feet of bridging material will be required for a job where 21 joists 2 by 10 on 12-in. centers are used.

Instructions	Computations
(a) Find the number of spaces. The number of spaces is one less than the number of joists. These are 20 spaces.	(a) $21 - 1 = 20$
(b) Multiply the number of spaces by 2 to get the number of pieces. There are 40 pieces.	(b) $20 \times 2 = 40$

(c) Multiply 14.142, the length of each piece, by 40 to get the total length.

(c) $14.142 \times 40 = 565.680$

(d) Divide by 12 to get the number of feet.

(d) $565.68 \div 12 = 47.14$

About 48 ft. of material will be required.

Problems	Work Here

(1) How many pieces of bridging are required for 17 joists if one row is used?

Answer:

(2) Determine the number of pieces for two rows of bridging if there are 29 joists.

Answer:

(3) Find the length of a piece of bridging for 2 by 6's on 12-in. centers.

Answer:

(4) What is the length of a piece of bridging for 2 by 8's on 12-in. centers?

Answer:

(5) Determine the length of a piece of bridging for 2 by 10's on 16-in. centers.

Answer:

Problems	Work Here

(6) How long is a piece of bridging
for 2 by 12's on 16-in. centers?

Answer:

(7) How many lineal feet of bridging
will be required for one row for
2 by 14's on 16-in. centers if
there are 15 joists?

Answer:

(8) Find the number of lineal feet of
bridging for two rows if 2 by 14's
on 12-in. centers are used.
There are 31 joists.

Answer:

(9) Two rows of bridging are required
for 25 joists on 12-in. centers.
The joists are 2 by 12's. How many
lineal feet of bridging are necessary?

Answer:

(10) How many lineal feet of bridging are
required for three rows of bridging
for 30 joists 2 by 12 on 12-in. centers?

Answer:

Subflooring

DO NOT SUBMIT THIS SELF-CHECK TEST FOR GRADING. Check your answers with the Answer Key at the end of this workbook.

Information

The subflooring, which is usually rough lumber of a cheaper variety, is laid after the joists are securely placed.

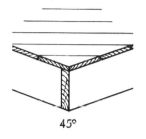 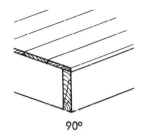 

45°   90°   PLYWOOD

Angle to Floor Joists

How To Find the Amount of Subflooring:

(1) Determine the entire floor area.
(2) Subtract for any openings such as stairways, etc.
(3) Add 30 per cent for boards applied diagonally.
(4) Add 20 per cent for boards not applied diagonally.
(5) Add 10 per cent for plywood.

Sample Problems

(1) Find the amount of subflooring for a floor 14 ft. by 22 ft. if it is applied diagonally.

$$A = 14 \times 22, \quad A = 308$$
$$308 \times .30 = 92.4$$
$$308 + 92.4 = 400.4$$

About 400 sq. ft. of subflooring are required.

(2) How much subflooring will be required for a floor 16 ft. by 34 ft. with an opening 5 ft. by 10 ft. in one corner using boards not applied diagonally?

$$A_1 = 16 \times 34, \quad A_1 = 544$$
$$A_2 = 5 \times 10, \quad A_2 = 50$$
$$A_1 - A_2 = 544 - 50, \quad A_1 - A_2 = 494$$
$$494 \times .20 = 98.8$$
$$494 + 98.8 = 592.8$$

About 593 sq. ft. of subflooring will be required.

(3) Calculate problem (2) above using plywood instead of boards.

$$494 \times .10 = 49.4$$
$$494 + 49.4 = 543.4$$

Problems	Work Here

(1) Determine the amount of subflooring
     needed for this floor using boards not
     applied diagonally.

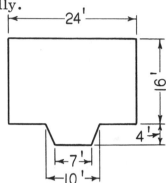

Answer:

(2) Find the amount of subflooring re-
     quired for this floor using diagonally
     applied boards.

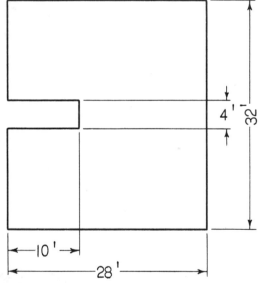

Answer:

(3-4) Using plywood subflooring, cal-
      culate the amount needed for
      problems 1 and 2 above.

Answer:

Answer:

Studs

DO NOT SUBMIT THIS SELF-CHECK TEST FOR GRADING. Check your answers with the Answer Key at the end of this workbook.

Information

Studs are vertical pieces usually 2 by 4's or 2 by 6's used in the framework of walls. Sheathing boards are nailed on the outside of the studs for exterior walls. For inside walls, lath are nailed to the studs. The length or height of the studs determines the height of the ceilings.

Two types of walls, the Western frame and the balloon frame, are used for most frame construction. Either kind of frame may have any type of surface such as siding, stucco, shingles, brick veneer, etc.

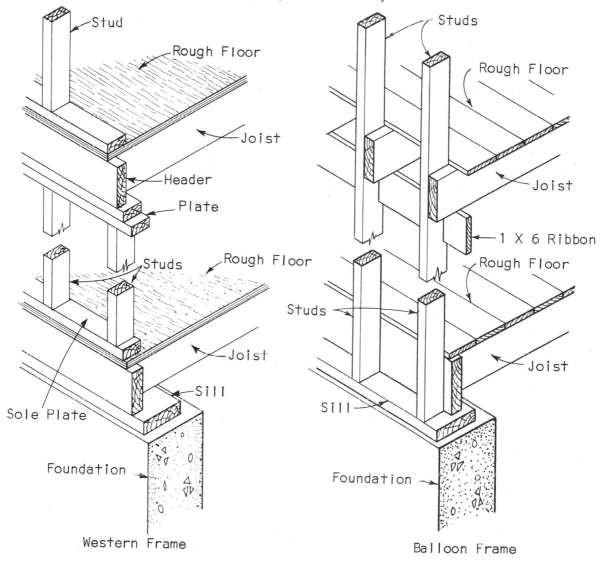

There are variations in the spacing of studs for exterior walls, but good practice calls for a spacing of 16 in. on center for the Western frame and 12 in. on center for the balloon frame. To allow for additional requirements in walls with openings 25 per cent is added to the estimate. Two sets of studs, one for each floor, are required for the Western frame.

How To Determine the Number of Studs

The number of studs required for any given perimeter (distance around) of a building may be estimated by

(1) determining the perimeter or distance around the building;
(2) finding a fraction by dividing 12 by the required center to center spacing,
(3) then multiplying the given perimeter by this fraction.

Sample Problem 1

How many studs will be required for a residence 28 ft. by 38 ft. if the balloon type frame with 16-in. spacing is used?

(1) Determine the perimeter of the building.

$$P = 2\,L + 2\,W = (2 \times 38) + (2 \times 28) = 76 + 56 = 132$$

The perimeter of the building is 132 ft.

(2) Find the fraction.

$$\frac{\overset{3}{\cancel{12}}}{\underset{4}{\cancel{16}}} = \frac{3}{4} \qquad \text{The fraction is } \frac{3}{4}.$$

(3) Multiply the perimeter by this fraction.

$$\overset{33}{\cancel{132}} \times \frac{3}{\underset{1}{\cancel{4}}} = 99$$

The number of studs required is 99.

Sample Problem 2

Find the number of studs for a Western type frame 26 ft. by 42 ft. if 18-in. spacing is used

(1) Determine the perimeter of the building.

$$P = 2L + 2W = (2 \times 42) + (2 \times 26) = 84 + 52 = 136$$

The perimeter of the building is 136 ft.

(2) Find the fraction.

$$\frac{\overset{2}{\cancel{12}}}{\underset{3}{\cancel{18}}} = \frac{2}{3} \qquad \text{The fraction is } \frac{2}{3}$$

(3) Multiply the perimeter by this fraction.

$$136 \times \frac{2}{3} = \frac{272}{3} = 90\frac{2}{3}$$

The number of studs for one floor is 91, for two floors is $91 \times 2 = 182$.

Sample Problem 3

Determine the number of studs for a wall 60 ft. long with spacings of (a) 18 in., (b) 20 in., and (c) 24 in., allowing 20 per cent for openings. Add 1 stud when figuring for one wall

(a) $\quad \dfrac{\overset{2}{\cancel{12}}}{\underset{3}{\cancel{18}}} = \dfrac{2}{3} \qquad \overset{20}{\cancel{60}} \times \dfrac{2}{3} = 40 \qquad 40 \times .20 = 8 \qquad 40 + 8 = 48$

The number of studs required is $48 + 1 = 49$.

(b) $\quad \dfrac{\overset{3}{\cancel{12}}}{\underset{5}{\cancel{20}}} = \dfrac{3}{5} \qquad \overset{12}{\cancel{60}} \times \dfrac{3}{\underset{1}{\cancel{5}}} = 36 \qquad 36 \times .20 = 7.2 \qquad 36 + 8 = 44$

The number of studs required is $44 + 1 = 45$

(c) $\quad \dfrac{\overset{1}{\cancel{12}}}{\underset{2}{\cancel{24}}} = \dfrac{1}{2} \qquad \overset{30}{\cancel{60}} \times \dfrac{1}{\underset{1}{\cancel{2}}} = 30 \qquad 30 \times .20 = 6 \qquad 30 + 6 = 36$

The number of studs required is $36 + 1 = 37$.

Problems	Work Here

(1) Find the number of studs for a building
26 ft. by 38 ft. if it is (a) a balloon
frame with 12-in. spacing, (b) a two-
floor Western frame with 16-in. spacing.
Allow 20 per cent for openings in each
case.

Answers:

(a)

(b)

(2) How many studs will be required for this
two-story Western type frame which is
irregular in space? Use 16-inch spacing
and allow 20 per cent.

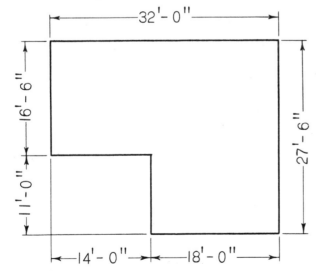

Answer:

Plates, Wall Sills, and Ribbons

DO NOT SUBMIT THIS SELF-CHECK TEST FOR GRADING. Check your answers with
the Answer Key at the end of this workbook.

Information

The horizontal member of a wall or partition frame to which the top end
of each stud is nailed are called <u>plates.</u>   The bottom ends are nailed to sills.
Two plates are used at the top of each wall or partition. Plates are usually
2 in. by 4 in. material.

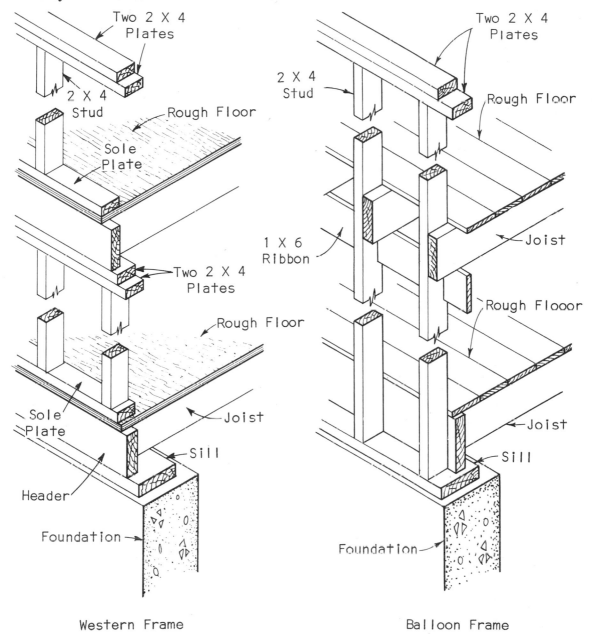

Western Frame   Balloon Frame

How To Determine the Number of Linear Feet of Plates

Usually two plates and one sill are required for each floor of the outside wall and each partition for the Western frame.

To find the number of linear feet of plate stock and sill stock, measure all walls and partitions for each floor. Multiply this linear measure by 3 to determine the number of linear feet of plate stock and sill stock.

Since the studding extends from the sill to the top plate of the second story in balloon framing it is only necessary to determine the linear feet of plate stock for the second story. However, to form a support for the second floor joists, ribbons are used. A ribbon is a piece of 1-inch material set into a notch in the studs of buildings of two stories.

To find the linear feet of plate material for the top plates of the balloon frame, multiply the perimeter of the building by 2. The number of linear feet of ribbon will be the same as the perimeter. Each partition wall for balloon framing will require 3 plates.

Sample Problem 1

Find the linear feet of outside wall plates and sills to be ordered for a two-story Western frame building which is 24 ft. wide and 34 ft. long.

Instructions	Computations
(1) Find the perimeter of the build- ing. The perimeter of the build- ing is 116 ft.	(1) P $= (2 \times L) + (2 \times w)$ $= (2 \times 34) + (2 \times 24)$ $= 68 + 48$ $= 116$
(2) Multiply this result by 3 to deter- mine the linear feet for 1 floor.	(2) $116 \times 3 = 348$
(3) Multiply the result thus obtained by 2 to find the amount for both stories.	(3) $348 \times 2 = 696$

The number of linear feet of plates and sills required for the job is 696.

Sample Problem 2

Determine the linear feet of top plates and ribbon for the exterior walls of a balloon frame type building 26 ft. wide and 34 ft. long.

<table>
<tr><td align="center">Instructions</td><td align="center">Computations</td></tr>
<tr>
<td>(1) Find the perimeter of the building. The perimeter of the building is 120 ft.</td>
<td>(1) $P = (2 \times L) + (2 \times w)$
$= (2 \times 34) + (2 \times 26)$
$= 68 + 52$
$= 120$</td>
</tr>
<tr>
<td>(2) Multiply this result by 2 to get the linear feet of plates.</td>
<td>(2) $120 \times 2 = 240$</td>
</tr>
<tr>
<td>(3) Multiply the perimeter by 1 to get the linear feet of ribbon.</td>
<td>(3) $120 \times 1 = 120$</td>
</tr>
</table>

This job requires 240 linear ft. of plates and 120 ft. of ribbons.

Sample Problem 3

Find the linear feet of plates and sills for the exterior walls of this two-story Western frame.

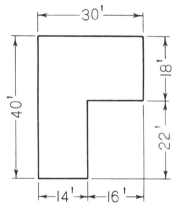

<table>
<tr><td align="center">Instructions</td><td align="center">Computations</td></tr>
<tr>
<td>(1) Find the perimeter of the building. The perimeter is 140 ft.</td>
<td>(1) $P = 40 + 30 + 18 + 16 + 22 + 14$
$= 140$</td>
</tr>
<tr>
<td>(2) Multiply this result by 3 to get the linear feet of plates and sills for one story.</td>
<td>(2) $140 \times 3 = 420$</td>
</tr>
<tr>
<td>(3) Multiply the linear feet of plate and sills for one story by 2 to get the linear feet for both stories.</td>
<td>(3) $420 \times 2 = 840$</td>
</tr>
</table>

This job requires 840 linear ft. of plates and sills.

Problems	Work Here

(1) Find the linear feet of plates
and sills required for the inside
walls of this building.

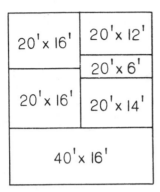

Answer:

(2) Find (a) the linear feet of plates
and sills for a two-story Western
frame,(b) the linear feet of plates
and ribbons for a two-story
balloon frame for this building.

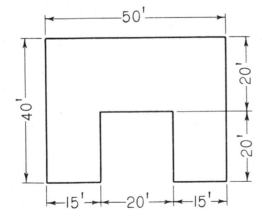

Answer:

Answer:

Sheathing

DO NOT SUBMIT THIS SELF-CHECK TEST FOR GRADING. Check your answers with the Answer Key at the end of this workbook.

Information

Plywood sheets 4 ft. by 8 ft. or sometimes wide boards 4 in. to 6 in. wide are nailed to studding to be used as a foundation for the covering of the exterior walls of a building. This is called <u>sheathing</u>. Shingles, clapboards, siding, and other surfacing material are nailed to the sheathing. Before nailing the outer surfacing, building paper is placed on the sheathing.

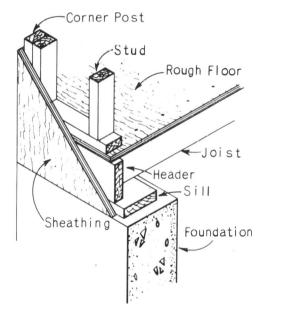

PLYWOOD SHEATING

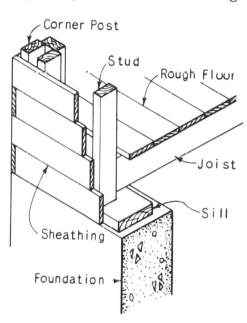

RIGHT ANGLE SHEATING

How To Determine the Amount of Material for Plywood Sheathing
    (1) Multiply distance from sill to top plates by the perimeter of the building.
    (2) Multiply the roof rise by one-half the span to find areas of gables.
    (3) Use these calculations. No waste allowance is required since cutouts from doors and windows may be used.

How To Determine the Amount of Material for Right Angle Sheathing
    (1) Multiply distance from sill to top plates by the perimeter of the building.
    (2) Subtract the areas of all openings such as doors and windows.
    (3) Add 30 per cent for waste
    (4) Multiply the roof rise by one-half the span to find the areas of gables.

Problems	Work Here

(1) One side of a building is 36 ft. long. The distance from the sill to the top plates is 10 ft. How much plywood sheathing is required for this wall?

Answer:

(2) How much sheathing will be required for a building 25 ft. long and 14 ft. wide if the distance from the sill to top plates is $9\frac{1}{2}$ ft.? There is one opening 5 ft. wide and 7 ft. high and two openings 2 ft. wide and 3 ft. 6 in. high. Allow 30 per cent waste.

Answer:

(3) Find the amount of sheathing for a building 24 ft. wide and 40 ft. long if the distance from the sill to the top plates is 14 ft. Deduct for 3 doors 3 ft. by 7 ft. and 12 windows 2 ft. 4 in. by 3 ft. 6 in. Allow 30 per cent waste.

Answer:

RELATED MATHEMATICS FOR CARPENTERS

Examination No. 9

Based on pages 165 to 180 in this workbook

Student's Name_____ Student Number_____

Street_____City_____State_____Code_____

Problems	Show Work Here

Note: There is one less space between joists than the number of joists, and each space uses two pieces of bridging.

(1) Find the number of pieces of bridging required for 19 joists if one row
is used.

Answer:

(2) How many pieces of bridging are
needed for two rows if there are
15 joists?

Answer:

(3) Determine the length of a piece
of bridging for 2 by 8's on 12-in.
centers.

Answer:

(4) Find the number of lineal feet for
two rows of bridging if 2 by 6's on
16-in. centers are used. There
are 25 joists.

Answer:

Problems	Show Work Here

(5) How many square feet of subflooring are needed for a floor $22\frac{1}{2}$ ft.
long and 14 ft. wide if it is applied
at right angles to the joists?

Answer:

(6) Plywood subflooring is applied to
a floor which is 24 ft. long and 12
ft. 4 in. wide. How much subflooring is needed?

Answer:

(7) There are two openings 12 ft. long
and $3\frac{1}{2}$ ft. wide in a floor $20\frac{1}{2}$ ft.
long and 16 ft. wide. Determine
the amount of subflooring if it is
applied diagonally.

Answer:

(8) A balloon type frame building is 42 ft.
long and 26 ft. wide. How many studs
are required if 16-in. spacing is used?

Answer:

(9) Determine the number of studs for a
Western type 2-story frame building
38 ft. long and $25\frac{1}{2}$ ft. wide if 16-in.
spacing is used.

Answer:

(10) A two-story Western frame building
is 37 ft. long and 22 ft. wide. Find
the number of lineal feet of outside
wall plates and sills required.

Answer:

Student's Name_____ Student Number_____

Problems	Show Work Here

(11-12) How many lineal feet of top
plates and ribbons are required
for the exterior walls of a balloon
type frame $32\frac{1}{2}$ ft. long and 22

ft. wide?
Answer:

Answer:

(13) Find the lineal feet of plates and
sills for the inside partitions of
this building.

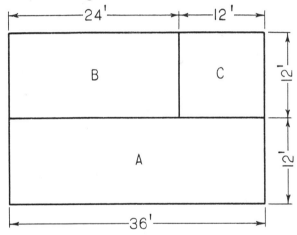

Answer:

(14-16) How many square feet of subflooring applied diagonally are required for each of the rooms A, B,
and C?

Room A
Answer:

Room B
Answer:

Room C
Answer:

Problems	Show Work Here

(17) Determine the number of square feet of subflooring applied at right angles for this room.

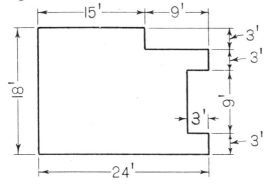

Answer:

(18-19) How many studs are needed for a wall 48 ft. long with spacings of (a) 16 in. and (b) 24 in. ?

Answer:

Answer:

(20) The distance from the sill to the top plates of a building 35 ft. long and 20 ft. wide is 12 ft. There are 2 doors 4 ft. wide and 7 ft. high and 10 windows $2\frac{1}{2}$ ft. wide and 4 ft. high. How much sheathing will be required? Allow 30 per cent waste.

Answer:

Roof Areas

Information

The shed, hip, gable, and gambrel are the styles of roofs in common use.
Determining the amounts of various materials involves finding the areas of
rectangles, triangles, and trapezoids.

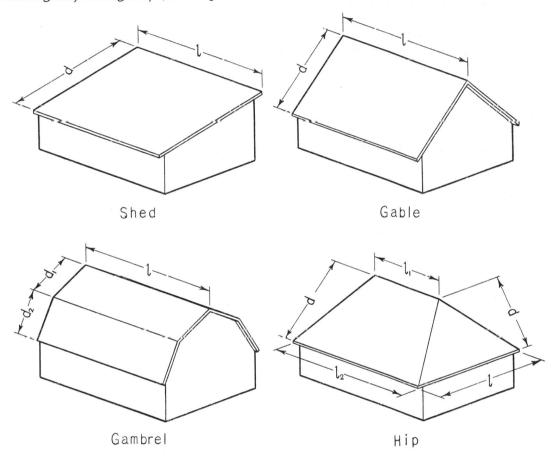

How To Find the Areas

(1) Shed Roof

To find the area of a shed roof, multiply the length of the roof (l) by the
length of the rafter (d).

$$A = l \times d \quad \text{or} \quad A = l \, d$$

(2) Gable Roof

To find the area of a plain gabled roof, multiply the length of the ridge ( )
by the length of the rafter (d), then multiply the product by 2.

$$A = l \times d \times 2 \quad \text{or} \quad A = 2 \times l \times d \quad \text{or} \quad A = 2 \, l \, d$$

(3) Gambrel Roof

To find the area of a gambrel roof, multiply the length of the ridge (l) by the sum of the widths of the two sections (d_1 and d_2), then multiply this product by 2.

$$A = l \times (d_1 + d_2) \times 2 \ \text{ or } \ A = 2 \times l \times (d_1 + d_2) \ \text{ or } \ A = 2\,l\,(d_1 + d_2)$$

(4) Equal-Pitch Hip Roof

To find the area of the triangular sections of an equal-pitch hip roof, multiply the length of the eave (l) by one--half the common rafter length d, then multiply by 2.

$$\text{Area of triangles} = l \times \frac{1}{2}d \times 2 \ \text{ or } \ l \times d \ \text{ or } \ l\,d$$

To find the area of the trapezoidal sections of an equal-pitch hip roof, multiply one-half the common rafter length d by the sum of the lengths of the ridge l_1 and the eaves l_2 , then multiply by 2

$$\text{Area of trapezoids} = \frac{1}{2}d \times (l_1 + l_2) \times 2 \ \text{ or } \ d \times (l_1 + l_2) \ \text{ or } \ d\,(l_1 + l_2)$$

$$\text{Total area of roof} \ = \ l\,d \ ^+ \ d(l_1 + l_2) \ \text{or} \ d \times (l + l_1 + l_2)$$

Sample Problems

(1) The rafter length of a shed roof 28 ft. long is $10\frac{1}{2}$ ft. Find the roof area.

$$A = l \times d$$

$$= 28 \times 10\frac{1}{2}$$

$$= 294$$

The area of the roof is 294 sq. ft.

(2) A gable roof 36 ft. long has rafters 14 ft. long. What is the area of the roof?

$$A = l \times d \times 2$$

$$= 36 \times 14 \times 2$$

$$= 1,008$$

The area of the roof is 1,008 sq. ft.

(3) The rafter lengths of the two sections of a gambrel roof which is 72 ft. long are 18 ft. and 12 ft. Find the area of the roof.

$$A = 2 \times l \times (d_1 + d_2)$$

$$= 2 \times 72 \times (18 + 12)$$

$$= 2 \times 72 \times 30$$

$$= 4,320$$

The area of the roof is 4,320 sq. ft.

(4) The length at the eaves of an equal-pitch hip roof is 30 ft., the span 18 feet, and the ridge 12 ft. Find the roof area if the common rafter length is 10 3/4 ft.

$$A = d \times (l + l_1 + l_2)$$

$$= 10.75 \times (18 + 12 + 30)$$

$$= 10.75 \times 60$$

$$= 645$$

The area of the roof is 645 sq. ft.

(5) How many square feet of roof boards are required for two sections of a roof, one of which is $18\frac{1}{2}$ ft. long and 14 ft. wide and the other with parallel sides 20 ft. and 28 ft. long with a width of $10\frac{1}{2}$ ft.?

$$A_1 = 18\frac{1}{2} \times 14$$

$$= 259$$

$$A_2 = \frac{10.5 \, (20 + 28)}{2}$$

$$= 252$$

$$A_1 + A_2 = 259 + 252$$

$$= 511$$

The area of the two roof sections is 511 sq. ft.

Problems	Work Here

(1) What is the roof area of a shed roof 42 ft. long if the rafter lengths are $16\frac{1}{2}$ ft.?

Answer:

(2) The rafter length of a gable roof 28 ft. long is 12 ft. Determine the roof area.

Answer:

(3) Determine the area of a gambrel roof 60 ft. long if the rafter lengths of the sections are $16\frac{1}{2}$ ft. and $10\frac{1}{2}$ ft.

Answer:

(4) An equal-pitch hip roof of a building has a length of 44 ft. at the eaves, a span of 28 ft. and a ridge length of 16 ft. Find its area if the common rafter length is 16.125 ft.

Answer:

(5) Find the area of this section of a roof:

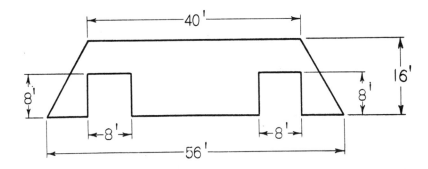

Answer:

Roof Pitches

DO NOT SUBMIT THIS SELF–CHECK TEST FOR GRADING. Check your answers with
the Answer Key at the end of this workbook.

Information

The <u>pitch of a roof</u> is the angle or slant of a roof from the plate to the
ridge.
The vertical distance which the rafters extend from the plate to the ridge
is called the <u>rise.</u>

The horizontal distance from the outer edge of the plate to a plumb line
dropped from the center of the ridge or highest point of the rafter is called the
<u>run.</u>

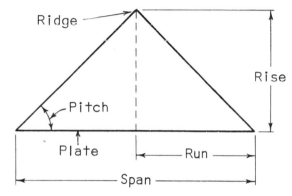

How To Figure Pitch, Rise, Run, or Span

The pitch of a roof is found by <u>dividing</u> the rise by the span.

$$\text{Pitch} = \frac{\text{Rise}}{\text{Span}}$$

The <u>span</u> is equal to <u>twice</u> the <u>run.</u> It is also equal to the <u>rise divided</u>
by the <u>pitch.</u>

$$\text{Span} = 2 \times \text{Run}$$

$$\text{Span} = \frac{\text{Rise}}{\text{Pitch}}$$

The <u>rise</u> is found by <u>multiplying</u> the <u>span</u> by the <u>pitch.</u>

$$\text{Rise} - \text{Span} \times \text{Pitch}$$

The <u>run</u> is equal to <u>one-half</u> the <u>span.</u> It is also equal to the <u>rise</u> divided
by <u>twice</u> the <u>pitch.</u>

$$\text{Run} = \frac{1}{2}\ \text{Span}$$

$$\text{Run} = \frac{\text{Rise}}{2 \times \text{Pitch}}$$

Sample Problem

What is the pitch of a roof if the span is 24 ft. and the rise 8 ft.?

$$\text{Pitch} = \frac{8}{24}\ \text{or}\ \frac{1}{3}$$

The pitch of the roof is $\frac{1}{3}$

Problems	Work Here

(1) Find the span of a roof if the pitch is $\frac{1}{4}$ and the rise 6 ft.

Answer:

(2) The span of a roof is 24 ft. and the pitch $\frac{1}{2}$. What is the rise?

Answer:

(3) The span of a roof is 26 ft. and the rise 8 ft. 8 in. Find the pitch.

Answer:

(4) What is the run if the rise is 18 ft. and the pitch $\frac{3}{4}$?

Answer:

Rafters

DO NOT SUBMIT THIS SELF–CHECK TEST FOR GRADING. Check your answers with the Answer Key at the end of this workbook.

Information

The sloping members of a roof extending from the ridge or hip of a roof to the eaves are called <u>rafters</u>. Rafters are purchased in multiples of 2-foot lengths.

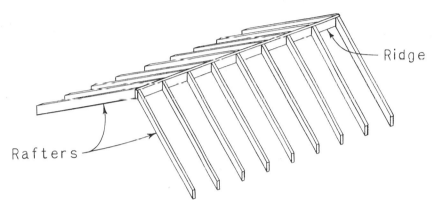

It is very important for carpenters to be able to determine various lengths and cuts for rafters. While it is true that these can be obtained by using the steel square, it is also well to understand the mathematics involved.

How To Use the Square Root Method To Determine Rafter Lengths

The square root method is based upon the relationship between the sides of a right triangle.

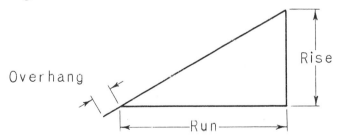

$$\text{Length of Rafter} = \sqrt{(\text{Rise})^2 + (\text{Run})^2} + \text{overhang}$$

Sample Problems

(1) Determine the length of rafters for a gable roof with a rise of 6 ft. and a run of 8 ft. if the overhang is 1 ft.

$$\text{Length of Rafter} = \sqrt{6^2 + 8^2} + 1$$
$$= \sqrt{36 + 64} + 1$$
$$= \sqrt{100} + 1$$
$$= 10 + 1$$
$$= 11$$

The rafters are 11 ft. long.

(2) A shed type roof has a rise of 4 ft. and a run of 14 ft. What is the length of rafter? Allow 1 ft. for overhang.

$$\text{Length of Rafter} = \sqrt{16 + 196} + 1$$
$$= \sqrt{212} + 1$$
$$= 14.56 + 1$$
$$= 15.56$$

The rafters are 15 ft. $6\frac{3}{4}$ in. long.

(3) The pitch of a gable roof is $\frac{1}{4}$. Find the length of rafters if the span is 32 ft. Allow 1 ft. for overhang.

$$\text{Rise} = \text{Span} \times \text{Pitch}$$
$$= 32 \times \frac{1}{4}$$
$$= 8$$
$$\text{Run} = \frac{1}{2} \text{ Span}$$
$$= \frac{1}{2} \times 32$$
$$= 16$$
$$\text{Length of Rafter} = \sqrt{8^2 + 16^2} + 1$$
$$= \sqrt{64 + 256} + 1$$
$$= \sqrt{320} + 1$$
$$= 17.889 + 1$$
$$= 18.889$$

The rafters are 18 ft. $10\frac{11}{16}$ in.

How To Find the Lengths of Hip Rafters

    The shape of either end of an equal-pitch hip roof is triangular. If the common rafter length and the span of the roof are known, then the square root method can be used to find the length of the hip rafter.

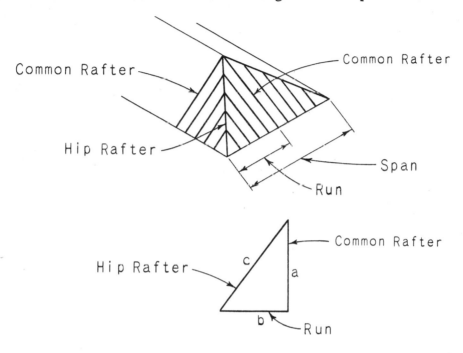

Length of Hip Rafter $= \sqrt{a^2 + b^2}$

Sample Problem

    The span of a building is 24 ft. and the common rafter length is 19 ft. What is the length of the hip rafter? Allow $1\frac{1}{2}$ ft. for overhang.

$$\text{Run} = \frac{1}{2} \times \text{Span}$$

$$= \frac{1}{2} \times 24$$

$$= 12$$

$$\text{Length of Hip Rafter} = \sqrt{19^2 + 12^2} + 1.5$$

$$= \sqrt{361 + 144} + 1.5$$

$$= \sqrt{505} + 1.5$$

$$= 22.47 + 1.5$$

$$= 23.97$$

    The length of the hip rafter is 23 ft. $11\frac{5}{8}$ in.

How To Find the Lengths of Valley Rafters

The square root method can be used to find the length of valley rafters if the span of the main building and the length of the common rafters of the wing are known.

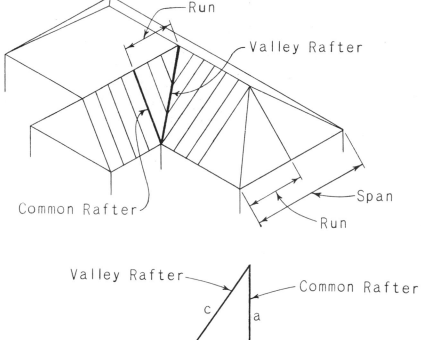

$$\text{Length of Valley Rafter} = \sqrt{a^2 + b^2}$$

Sample Problem

The span of the main part of a building is 40 ft. The length of the common rafters of the wing having the same ridge height are 25 ft. Find the length of the valley rafter. Allow 1 ft. overhang.

$$\text{Run} = \frac{1}{2} \times \text{Span}$$

$$= \frac{1}{2} \times 40$$

$$= 20$$

$$\text{Length of Valley Rafter} = \sqrt{25^2 + 20^2} + 1$$

$$= \sqrt{625 + 400} + 1$$

$$= \sqrt{1025} + 1$$

$$= 32.01 + 1$$

$$= 33$$

The length of the valley rafter is 33 ft.

How To Find the Number of Roof Rafters for Shed, Gable, and Gambrel Roofs

For simple roofs such as the shed and gable, finding the number of rafters is rather easy. This is also true of the gambrel roof since there are two rectangular sections on each side.

(1) If the rafters are spaced 16 in. on center, multiply the length of the roof by .75 and then add 1.

(2) If the rafters are spaced 12 in. on center, allow one rafter for each foot and then add 1.

Sample Problems

(1) Find the number of rafters for a shed type roof 32 ft. long if they are spaced 16 in. on center.

$$(32 \times .75) + 1 = 24 + 1 = 25$$

The number of rafters required is 25.

(2) How many rafters will be required for a gable roof 36 ft. long if the rafters are spaced 16 in. on center?

$$(36 \times .75) + 1 = 27 + 1 = 28 \text{ (one side)}$$
$$28 \times 2 = 56 \text{ (both sides)}$$

The number of rafters required is 56.

(3) Determine the number of rafters for a gambrel roof whose length is 40 ft. Allow one set for each section of each side.

$$(40 \times .75) + 1 = 30 + 1 = 31 \text{ (one section)}$$
$$31 \times 2 = 62 \text{ (two sections on one side)}$$
$$62 \times 2 = 124 \text{ (four sections for two sides)}$$

The number of rafters required is 124.

Since a gable roof and a hip roof having the same dimensions and pitch require the same amount of material, the number of rafters can be determined in the same way as for the gable roof.

Problems	Work Here

(1) The rise and run of a shed roof are $3\frac{1}{2}$ ft. and 20 ft. Find the length of rafter. Allow $1\frac{1}{2}$ ft. for overhang.

Answer:

(2) What is the length of rafter for a gable roof with a rise of 12 ft. and a run of 16 ft. ? Allow 1 ft. for overhang.

Answer:

(3) The pitch of a gable roof is $\frac{1}{2}$. Find the length of rafter if the span is 30 ft. Allow 1 ft. for overhang.

Answer:

Problems	Work Here

(4) How long is a hip rafter of an equal-
pitch roof if the common rafter length
is 21 ft. and the span is 38 ft.? Allow
1 1/4 ft. overhang.

Answer:

(5-6) Find the length of the lower and
upper rafters of a gambrel roof
if the rise and run of the lower
are 18 ft. and 6 ft. and the upper
are 5 ft. and 15 ft. Allow 1 ft. for
overhang for lower rafters.

Answer:

Answer:

(7) The length of the common rafters
of the wing of a building is 20 ft.
The span of the main building is
36 ft. Find the length of valley
rafter if both parts have the same
ridge height. Allow 1 ft. overhang.

Answer:

Problems	Work Here

(8) How many rafters are required for a shed type roof 56 ft. long if they are spaced 16 in. on center?

Answer:

(9) Find the number of rafters for a gambrel roof 44 ft. long if the spacing is 12 in. on center.

Answer:

(10) How many rafters are required for a gable roof 52 ft. long if the spacing is 16 in. on center?

Answer:

Siding

DO NOT SUBMIT THIS SELF-CHECK TEST FOR GRADING. Check your answers with the Answer Key at the end of this workbook.

Information

   Exterior surfacing such as siding, clapboard, shingles, etc., are nailed to the sheathing with building paper between. Boards used for this purpose are called <u>siding</u>.

   A board which is thicker along one of its edges is called <u>bevel siding.</u> The thicker edge is placed so that it overlaps the thinner edge of the siding which is below it.

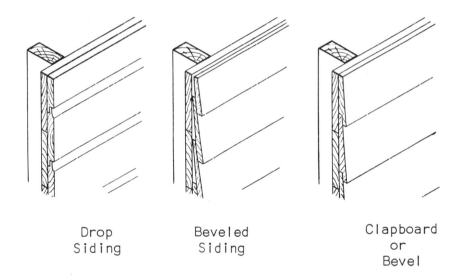

| Drop Siding | Beveled Siding | Clapboard or Bevel |

How To Determine the Amount of Siding

   Practice in estimating the waste, overlapping, and allowance for windows, doors, etc., in applying siding varies considerably. Undoubtedly, no two estimates will agree.

   In general, the procedure for estimating the amount of bevel or lap siding is as follows:

   (1) Determine the total area in square feet of all surfaces to be covered with siding.

   (2) Deduct for all openings over 10 sq. ft. to obtain area to be covered by siding.

   (3) Multiply this area by a constant found in the following table.

TABLE OF CONSTANTS

Type	Size	Lap	Wall	
			Plain	Openings
Bevel	1 × 4	$\frac{3}{4}$ in.	1.45	1.49
Bevel	1 × 6	1 in.	1.39	1.43
Bevel	1 × 8	$1\frac{1}{4}$ in.	1.33	1.37
Bevel	1 × 10	$1\frac{1}{4}$ in.	1.29	1.33
Bevel	1 × 12	$1\frac{1}{2}$ in.	1.23	1.27
Drop	1 × 4	Shiplap	1.28	1.32
Drop	1 × 6	Shiplap	1.19	1.23
Drop	1 × 8	Shiplap	1.16	1.20

Sample Problems

(1) Find the number of square feet of siding required for a plain wall if the height is 15 ft. and the length 32 ft. if 1 by 6 siding is used with a 1 in. lap.

$$Area = 15 \times 32$$
$$= 480$$
$$480 \times 1.39 = 667.2$$

The number of square feet of siding is 667.

(2) How much 1 by 8 siding with a $1\frac{1}{4}$ in. lap will be required for a wall 14 ft. high and 24 ft. long if there are two windows 3 ft. by 4 ft. 6 in. and one door 3 ft. by 7 ft.?

$$Area\ of\ wall = 14 \times 24$$
$$= 336$$
$$Area\ of\ windows = 2 \times 3 \times 4\frac{1}{2}$$
$$= 27$$
$$Area\ of\ door = 3 \times 7$$
$$= 21$$

$$Area\ of\ wall - Area\ of\ openings = 336 - (27 + 21)$$
$$= 336 - 48$$
$$= 288$$
$$288 \times 1.37 = 394.56$$

The number of square feet of siding is 395.

How To Determine the Amount of Shingles for Side Walls

Usually shingles are purchased by the square or in bundles. There are four bundles in one square which equals 100 sq. ft.

Surface Covered by One Square of Shingles

Exposure Inches	Surface Covered in Square Feet		
	16 in. Shingles	18 in. Shingles	24 in. Shingles
$5\frac{1}{2}$	113	....	....
6	123	111	....
$6\frac{1}{2}$	134	120	....
7	144	129	....
$7\frac{1}{2}$	154	139	....
8	....	148	108
$8\frac{1}{2}$	....	157	115
9	....	....	122
$9\frac{1}{2}$	....	....	128
10	....	....	135
$10\frac{1}{2}$	....	....	142
11	....	....	155

(1) To find the number of squares of shingles required to cover a wall surface, find the area of the wall and then divide by the number indicating the surface covered for the given exposure.

(2) To find the number of bundles, multiply the number of squares by 4.

Sample Problem

How many squares and bundles of 18-inch shingles laid $6\frac{1}{2}$ in. to the weather will be needed to cover a plain wall 15 ft. high and 30 ft. long?

$$\text{Area of wall} = 30 \times 15$$
$$= 450$$
$$450 \div 120 = 3.75$$

The number of squares is $3\frac{3}{4}$.

$$3\frac{3}{4} \times 4 = 15$$

The number of bundles is 15.

Problems	Work Here

(1) How many square feet of bevel siding are required for a plain wall $28\frac{1}{2}$ ft. long and 14 ft. high if 1 by 8 siding is used with a $1\frac{1}{4}$ - in. lap?

Answer:

(2) A wall 20 ft. long and $9\frac{1}{2}$ ft. high has one opening 4 ft. wide and 7 ft. high. How much drop siding 1 by 6 is required?

Answer:

(3) A wall 34 ft. long and 15 ft. high has 4 windows 2 ft. by 3 ft. 10 in. and one door 3 ft. by 7 ft. How much bevel siding 1 by 4 with a $\frac{3}{4}$ - inch lap is required?

Answer:

(4) Find the number of squares and bundles of 16-in. shingles laid $7\frac{1}{2}$ in. to the weather which will be needed to cover a plain wall $42\frac{1}{2}$ ft. long and 16 ft. high.

Answer:

(5) How many bundles of 24-in. shingles laid 8 in. to the weather are needed for a wall 48 ft. long and 16 ft. high if there are two openings 4 ft. wide and 8 ft. high and four openings 2 ft. wide and $3\frac{1}{2}$ ft. high?

Answer:

Shingles

DO NOT SUBMIT THIS SELF-CHECK TEST FOR GRADING. Check your answers with the Answer Key at the end of this workbook.

Information

The lengths of shingles are 16 in., 18 in., or 24 in. The standard thicknesses are 4/2, $5/2\frac{1}{2}$ and 5/2. The meaning of 4/2 is 4 shingles to 2 in. of butt thickness. Likewise $5/2\frac{1}{2}$ and 5/2 mean 5 shingles to $2\frac{1}{2}$ in. of butt thickness and 5 shingles to 2 in. of butt thickness.

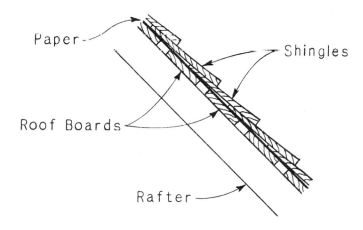

Surface Covered by One Square of Shingles

Exposure Inches	Surface Covered in Square Feet		
	16 in. Shingles	18 in. Shingles	24 in. Shingles
4	82	58	....
$4\frac{1}{2}$	92	83	....
5	103	92	....
$5\frac{1}{2}$	....	102	....
6	....	....	81
$6\frac{1}{2}$	....	....	88
7	....	....	95
$7\frac{1}{2}$	....	....	101

Problems	Work Here

(1-2) How many squares and bundles
of 24-in. shingles laid $6\frac{1}{2}$ in. to
the weather will be needed for a
gable roof if the ridge length is
25 ft. and the rafter length is
16 ft.? Allow 8 per cent waste.

Answer:

Answer:

(3) Determine the number of bundles
of 16-in. shingles laid 5 in. to the
weather for one side of a hip roof
if the length of the ridge is 24 ft.
and the length along the eaves 38
ft. The distance between the
parallel sides is 14 ft. Allow 12
per cent waste.

Answer:

(4) Find the number of squares of 24-in.
shingles exposed 6 in. to the weather
for the triangular end of a hip roof
if the altitude of the triangle is 14 ft.
and the length along the eaves 36.5 ft.
Allow 12 per cent waste.

Answer:

(5) How many bundles of 16-in. shingles
laid $4\frac{1}{2}$ in. to the weather will be required for this roof? Allow 8 per
cent waste.

Answer:

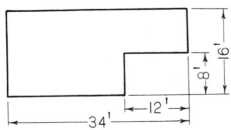

Finish Flooring

DO NOT SUBMIT THIS SELF–CHECK TEST FOR GRADING. Check your answers with the Answer Key at the end of this workbook.

Information

Finish flooring is nailed to the rough flooring which is covered with building paper.

If the finish flooring is to be covered with linoleum or carpeting, less expensive lumber such as pine may be used. Usually living rooms have oak, maple, etc., for finish flooring.

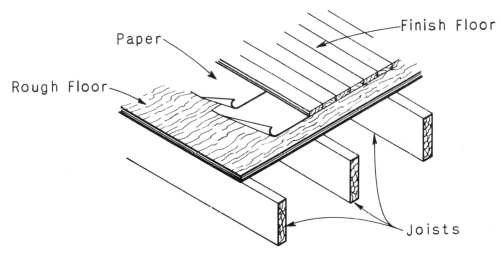

How To Determine the Amount of Finish Flooring

Find the area of each floor requiring the finish flooring. If $\frac{3}{4}$ in. by 2 in. or $\frac{3}{4}$ in. by $2\frac{1}{4}$ in. is used, add 33 per cent to the total area of all surface. Add 20 per cent to the total area of surface if 1 in. by 4 in. fir or pine is used.

Sample Problem

How many square feet of finish flooring are needed for a floor 14 ft. long and 10 ft. wide where $\frac{3}{4}$ in. by 2 in. flooring is used?

$$14 \times 10 = 140$$
$$140 \times .33 = 46.2$$
$$140 + 46.2 = 186$$

The number of square feet of finish flooring is 186 sq. ft.

Problems	Work Here

(1) Determine the number of square feet of $\frac{3}{4}$ in. by $2\frac{1}{4}$ in. finish flooring needed for a floor 24 ft. long and $15\frac{1}{2}$ ft. wide.

Answer:

(2) How many square feet of 1 in. by 4 in. pine flooring are required for a floor 22 ft. long and 14 ft. wide?

Answer:

(3-4) Find the number of square feet of $\frac{3}{4}$ in. by 2 in. flooring or 1 in. by 4 in. pine flooring needed for this floor.

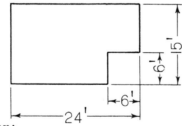

Answer:

Answer:

(5) How much more $\frac{3}{4}$ in. by $2\frac{1}{4}$ in. flooring will be needed than 1 in. by 4 in. pine flooring for this room?

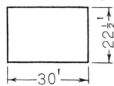

Answer:

Stairs

DO NOT SUBMIT THIS SELF-CHECK TEST FOR GRADING. Check your answers with the Answer Key at the end of this workbook.

Information

Finished stairs are usually made away from the job in a mill or shop. The stair is received on the job and the various parts such as the stringers, treads, risers, and railings are assembled. However, it may be necessary to construct basement and attic stairs.

The stair must be of sufficient width to allow the movement of objects such as furniture and appliances. It should also provide for comfortable stepping.

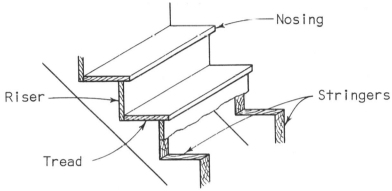

How To Find the Height of Risers and Width of Treads

The treads are one less in number than the risers

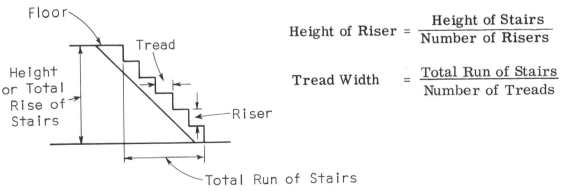

$$\text{Height of Riser} = \frac{\text{Height of Stairs}}{\text{Number of Risers}}$$

$$\text{Tread Width} = \frac{\text{Total Run of Stairs}}{\text{Number of Treads}}$$

Sample Problems

(1) The total rise of a flight of stairs is 12 ft. and an 8-inch riser is desired. Find the number of risers.

$$12 \text{ ft.} = 144 \text{ in.}$$

$$144 \div 8 = 18$$

There are 18 risers.

(2) Find the number of risers and height of each riser if the total rise of the stairs is 9 ft. 3 in. and a $7\frac{1}{8}$- in. riser is desired.

$$9 \text{ ft. } 3 \text{ in. } = 111 \text{ in.}$$

$$111 \div 7.125 = 15\frac{11}{19}$$

Since a part of a riser is not desirable, use 16 risers.

$$111 \div 16 = 6.9375$$

To the nearest eighth of an inch 7.4 in. equals $7\frac{7}{8}$ in. Use $7\frac{7}{8}$ in. as the height of each riser.

(3) What is the tread width for a stair with a run of 11 ft. 8 in. if the number of risers is 15?

$$\text{Number of treads } = 15 - 1 \text{ or } 14$$

$$11 \text{ ft. } 8 \text{ in. } = 140 \text{ in.}$$

$$140 \div 14 = 10$$

The width of each tread is 10 in.

(4) Find the width of the treads for a stairway with 14 risers and a total run of 10 ft. 10 in. if 1 in. per tread is allowed for nosing.

$$\text{Number of treads } = 14 - 1 \text{ or } 13$$

$$10 \text{ ft. } 10 \text{ in. } = 130 \text{ in.}$$

$$130 \div 13 = 10$$

$$10 + 1 = 11$$

The width of each tread is 11 in.

Problems	Work Here

(1) Find the number of risers for a
flight of stairs with a total rise
of 14 ft. 9 in. if a $7\frac{3}{8}$-in. riser
is desired.

Answer:

(2) The total rise of a flight of stairs
is 9 ft. 6 in. Find the number of
risers if the height of each is
$7\frac{1}{8}$ in.

Answer:

(3) How many $7\frac{11}{16}$-inch risers will
there be if the total rise of the
flight of stairs is 10 ft. 3 in. ?

Answer:

(4) What is the tread width for a stair
with a total run of 10 ft. if there
are 13 risers ?

Answer:

(5) The total run of a flight of stairs is
12 ft. 3 in. The number of risers
is 15. What is the width of the
treads ?

Answer:

Problems	Work Here

(6) What is the width of each tread for a stairway with 16 risers and a total run of 12 ft. 6 in. if a total of 15 in. is allowed for nosing?

Answer:

(7-8) What is the actual height of each riser and the number of risers if the total rise of the stairs is 9 ft. 6 in. and a $7\frac{3}{4}$-in. riser is desired?

Answer:

Answer:

(9) Find the width of treads for a flight of stairs with a total run of 11 ft. $10\frac{1}{2}$ in. and 16 risers if 1 in. per tread is allowed for nosing.

Answer:

(10) An 8-in. riser is desired for a stair with a total rise of 10 ft. 10 in. Find the actual height of each riser.

Answer:

RELATED MATHEMATICS FOR CARPENTERS

Examination No. 10

Based on pages 185 to 210 in this workbook

Student's Name_____Student Number_____

Street_____City_____State_____Zip Code_____

Problems	Show Work Here

(1) The rise of a roof is $7\frac{1}{2}$ ft. and the pitch is $\frac{1}{3}$. What is the span?

Answer:

(2) The span of a roof is 41 ft. and the rise 10 ft. 3 in. Find the pitch.

Answer:

(3) A gable roof has a rise of 8 ft. and a run of 12 ft. Find the rafter length allowing 1 ft. for overhang.

Answer:

(4) A gable roof has a pitch of $\frac{1}{3}$. What is the rafter length if the span is 39 ft.? Allow $1\frac{1}{2}$ ft. for overhang.

Answer:

(5) The common rafter of the triangular end of a hip roof is 10 ft. The span of the building is 28 ft. What is the length of the hip rafter, allowing $1\frac{1}{2}$ ft. for overhang?

Answer:

Problems	Show Work Here

(6) The length of the rafters of a wing having the same ridge height as the main part of the building is 14 ft. The span of the main part is 44 ft. Find the length of the valley rafter allowing 1 ft. for overhang.

Answer:

(7) How many rafters spaced 12 in. on center are required for a shed type roof 50 ft. long?

Answer:

(8) Find the roof area of a shed roof 60 ft. long if the rafter lengths are $14\frac{1}{2}$ ft.

Answer:

(9) The rafter lengths of the sections of a gambrel roof are 15 ft. and 12 ft. The ridge is 48 ft. long. Determine the entire roof area.

Answer:

(10) The ridge of a hip roof is 24 ft. long. The distance along the eaves is 38 ft. Find the roof area of <u>one</u> side of the roof if the rafter length is 13 ft.

Answer:

Student's Name_____Student Number_____

Problems	Show Work Here

(11) Find the number of square feet of bevel siding for a plain wall 24 ft. long and $11\frac{1}{2}$ ft. high if 1 by 10 siding with a $1\frac{1}{4}$-in. lap is used.

Answer:

(12) How much drop siding 1 by 8 is required for a wall 36 ft. long and $12\frac{1}{2}$ ft. high if there are 2 openings 4 ft. by 8 ft. and 6 openings 2 ft. 3 in. by 4 ft.?

Answer:

(13) Determine the number of squares of 16-in. shingles laid 7 in. to the weather required to cover a plain wall 30 ft. long and 11 ft. high. Allow 8 per cent waste.

Answer:

(14) Determine the number of bundles of 18-in. shingles laid $6\frac{1}{2}$ in. to the weather for a wall 44 ft. long and 14 ft. high if there are 3 openings 4 ft. wide and 9 ft. high and 6 openings 3 ft. wide and $3\frac{1}{2}$ ft. high. Allow 8 per cent waste.

Answer:

Problems	Show Work Here

(15) The ridge of a hip roof is 30 ft. long. The parallel distance along the eaves is 40 ft. The width of this roof section is 15 ft. Find the number of bundles of 18-in. shingles laid $4\frac{1}{2}$ in. to the weather required to cover it. Allow 12 per cent waste.

Answer:

(16) How many square feet of $\frac{3}{4}$ in. by 2 in. flooring are required for a floor 36 ft. long and $18\frac{1}{2}$ ft. wide?

Answer:

(17) The total run of a flight of stairs is 13 ft. 9 in. The number of risers is 16. What is the width of the treads? Allow 1 in. for nosing.

Answer:

(18) Find the actual height of each riser and the number of risers if the total rise of the stairs is 12 ft. and a $7\frac{1}{2}$ in. riser is required.

Answer:

Answer:

ANSWER KEY FOR RELATED MATHEMATICS FOR CARPENTERS

Self-Check Test 1, page 1
(1) 196 in. (2) 1,945 in. (3) 4,988 sq. in. (4) 6,082 cu. in. (5) 3,703 yd. (6) 7,629 sq. ft. (7) 7,355 sq. yd. (8) 9,762 (9) 14,343 lb. (10) 10,735 gal. (11) 58,081 cu. yd. (12) 98,344 B.t.u.

Self-Check Test 1, page 2
(1) 676 sq. ft. (2) 984 sq. ft. (3) 3,603 sq. in. (4) $7,734 (5) 65,075 cu. yd.

Self-Check Test 2, page 3
(1) 74 in. (2) 978 ft. (3) 227 yd. (4) 1,091 sq. in. (5) 2,591 sq. ft. (6) 2,085 sq. yd. (7) 5,857 cu. yd. (8) 12,588 lb. (9) 13,050 cu. ft. (10) $4,552 (11) 20,879 B.t.u. (12) $13,099

Self-Check Test 2, page 4
(1) 33 in. (2) 3,125 sq. ft. (3) 1,903 bricks (4) 5,415 sq. ft. (5) 52,125 cu. yd.

Self-Check Test 3, page 5
(1) 504 in. (2) 780 ft. (3) 2,728 sq. in. (4) 18,525 sq. ft. (5) 28,314 cu. in. (6) 255,088 sq. yd. (7) 152,205 cu. in. (8) 158,400 hr. (9) $67,500 (10) 183,168 cu. in. (11) 1,093,750 gal. (12) $1,237,500

Self-Check Test 3, page 6
(1) 588 in. (2) 1,100 ft. (3) 520 ft. (4) 1,312 shingles (5) $36,675

Self-Check Test 4, pages 7 and 8
(1) 108 (2) 311 (3) 3,071 (4) 576 (5) 96 (6) 351 (7) 32 (8) 102 (9) 1,058, remainder 138
(1) 6 pieces (2) $9\frac{1}{2}$ hr. (3) 125 hr. (4) 25 hr. (5) 13,575 B.T.U.

Self-Check Test 5, pages 9 and 10
(1) $\frac{3}{4}$ in. (2) $1\frac{1}{8}$ in. (3) $1\frac{5}{8}$ in. (4) $\frac{11}{12}$ yd. (5) $2\frac{1}{24}$ sq. ft. (6) $1\frac{11}{16}$ in. (7) $1\frac{1}{8}$ in.
(8) $1\frac{1}{16}$ (9) $\frac{7}{16}$ (10) $1\frac{5}{32}$ in.

Self-Check Test 6, pages 11 and 12
(1) $\frac{1}{4}$ in. (2) $1\frac{1}{8}$ in. (3) $\frac{1}{8}$ in. (4) $1\frac{1}{8}$ sq. ft. (5) $\frac{2}{3}$ sq. yd. (6) $\frac{3}{8}$ sq. yd. (7) $\frac{11}{16}$ in. (8) $\frac{11}{16}$ in.
(9) $\frac{1}{32}$ in. (10) $\frac{7}{16}$ in.

Self-Check Test 7, pages 13 and 14
(1) $9\frac{1}{2}$ in. (2) $5\frac{1}{4}$ ft. (3) 29 sq. in. (4) 34 yd. (5) $9\frac{3}{4}$ hr. (6) $3\frac{1}{8}$ in. (7) $6\frac{15}{16}$ in.
(8) $5\frac{1}{8}$ in. (9) $1\frac{5}{8}$ in. (10) $2\frac{7}{16}$ in.

Self-Check Test 8, pages 15 and 16
(1) $1\frac{1}{4}$ in. (2) $\frac{7}{16}$ in. (3) $\frac{13}{32}$ in. (4) $\frac{3}{32}$ in. (5) $\frac{1}{32}$ in. (6) $\frac{3}{4}$ in. (7) $\frac{3}{4}$ in. (8) $1\frac{2}{3}$ board feet (9) $\frac{9}{16}$ in. (10) $1\frac{5}{8}$ in.

Self-Check Test 9, pages 17 and 18
(1) $\frac{3}{8}$ in. (2) $\frac{9}{32}$ in. (3) $\frac{9}{16}$ in. (4) $\frac{3}{8}$ ft. (5) $\frac{1}{2}$ ft. (6) $\frac{1}{4}$ gal. (7) $\frac{1}{4}$ gal. (8) $\frac{1}{2}$ gal.
(9) $\frac{1}{2}$ gal. (10) 1 qt.

Self-Check Test 10, pages 19 and 20
(1) $1\frac{1}{4}$ in. (2) $\frac{1}{4}$ ft. (3) $1\frac{2}{3}$ sq. yd. (4) $\frac{5}{6}$ in. (5) $1\frac{1}{10}$ (6) $1\frac{3}{4}$ ft. (7) 4 pieces
(8) $3\frac{1}{2}$ ft. (9) $7\frac{1}{2}$ ft. (10) $5\frac{1}{2}$ ft.

Self-Check Test 11, pages 21 and 22
(1) $\frac{3}{16}$ in. (2) $\frac{1}{3}$ (3) 9 in. (4) $15\frac{1}{8}$ yd. (5) $35\frac{5}{9}$ ft. (6) $6\frac{3}{4}$ hr. or 6 hr. 45 min. (7) $4\frac{1}{8}$ in. (8) $6\frac{3}{4}$ hr. or 6 hr. 15 min. (9) $15\frac{1}{12}$ ft. or 15 ft. 1 in.
(10) $8\frac{1}{2}$ ft. or 8 ft. 6 in.

Self-Check Test 12, pages 23 and 24
(1) $\frac{1}{4}$ in. (2) $1\frac{1}{2}$ in. (3) $1\frac{1}{4}$ ft. (4) $1\frac{1}{4}$ in. (5) $37\frac{1}{2}$ sq. yd. (6) 4 shelf boards
(7) 12 pieces (8) 24 boards (9) 24 columns (10) 272 pieces

Self-Check Test 13, pages 29 and 30
(1) .625 in. (2) 1.5 in. (3) 16.5 in. (4) 1.625 in. (5) $174.23 (6) 2.0625 in.
(7) .875 in. (8) 6.625 in. (9) $2,936.96 (10) $86,161.44

Self-Check Test 14, pages 31 and 32
(1) 1.25 in. (2) .25 in. (3) 1.125 in. (4) 12.9 ft. (5) $25.38 (6) .035 in.
(7) 159.4 ft. (8) .31 lb. (9) $2,291.08 (10) $62.30

Self-Check Test 15, pages 33 and 34
(1) 3.75 in. (2) 3.125 in. (3) 5.9375 in. (4) 19.6875 in. (5) $2,067.06
(6) 580 lb. (7) 1,348.5 lb. (8) 1,267.2 lb. (9) 58.86 lb. (10) 151 bricks

Self-Check Test 16, pages 35 and 36
(1) 1.5 in. (2) 20 in. (3) 7 in. (4) 90 in. (5) $29.80 (6) 54 pieces (7) 24 table tops (8) 40 hr. (9) 72 hr. (10) $5.75

Self-Check Test 17, pages 37 and 38
(1) .750 (2) .625 (3) .1875 (4) .78125 (5) .703125 (6) 2.75 lb. (7) 1.375 in. (8) .1875 in. (9) .8125 in. (10) .34375 in.

Self-Check Test 18, pages 39 and 40
(1) $\frac{1}{4}$ (2) $\frac{7}{8}$ (3) $\frac{5}{8}$ (4) $\frac{7}{16}$ (5) $\frac{5}{8}$ (6) $\frac{11}{16}$ (7) $\frac{3}{8}$ in. (8) $\frac{5}{8}$ in. (9) $\frac{3}{16}$ in. (10) $\frac{1}{32}$ in.

ANSWER KEY FOR RELATED MATHEMATICS FOR CARPENTERS—Continued

Self-Check Test 19, page 42
(1) 6470 (2) 65.9 sq. ft. (3) 27.3 (4) 21790 (5) 272.2

Self-Check Test 20, pages 47 and 48
(1) 1,272 sq. ft. (2) 70 lb. (3) 99 lb. (4) 487.5 sq. ft. (5) 1,400 bricks

Self-Check Test 21, pages 49 to 52
(1) 33 in. (2) $4\frac{1}{2}$ ft. (3) $2\frac{2}{3}$ yd. (4) $8\frac{1}{2}$ sq. ft. (5) 96 sq. ft. (6) 160 cu. yd. (7) 101.5 cu. yd. (8) 22.44 gal. (9) 781.25 lb. (10) 1.225 tons (11) 73 in. (12) $55\frac{1}{2}$ sq. ft. (13) 16 ft. $2\frac{1}{2}$ in. (14) 5 ft. 3 in. (15) 198 cu. yd. 12 cu. ft. (16) 19 ft. 9 in. (17) 3 hr. 9 min. (18) 25° (19) 239° (20) 3.81 cm x 8.89 cm

Self-Check Test 22, pages 57 and 58
(1) 24.14 km (2) 2.1 gal. (3) 32 ft. $8\frac{1}{4}$ inches (4) 22,982.238 mm (5) 43,787.568 mm (6) 9 ft. 9 in. (7) 12 ft. $6\frac{1}{2}$ in. (8) 8 ft. (9) $37\frac{1}{2}$°C. (10) 68°F.

Self-Check Test 23, page 60
(1) 324 (2) 15.625 (3) 47.265625 (4) 307.546875 (5) $\frac{27}{64}$

Self-Check Test 23, page 62
(1) 25 (2) 91 (3) 1.07 (4) 11.2249 or 11.225 (5) 2.30

Self-Check Test 24, pages 63 and 64
(1) 2:3 (2) 7:5 (3) 2:3 (4) 4:5 (5) 2:5 (6) 9:32 (7) 15:32 (8) 3:4 (9) 1:2 (10) 1:4

Self-Check Test 25, pages 65 and 67
(1) 16 in. (2) 30 ft. (3) 7 (4) 10 (5) 440 cu. yd. (6) 22.5 lb. (7) 16.425 lb. (8) 7.92 cu. yd. (9) 4.5 gal. (10) 1,625 bricks

Self-Check Test 26, pages 68 and 69
(1) 4 x s (2) (2 x b) + (2 x h) (3) π x d (4) 2 x π x r (5) 1.414 x s (6) s x s (7) b x h (8) π x r x r (9) $\frac{1}{2}$ x b x h (10) .433 x s x s (11) 2.598 x s x s (12) 4 x π x r x r (13) e x e x e (14) .7854 x d x d x h (15) .5236 x d x d x d (16) .2618 x d x d x h (17) $\sqrt{(a \times a) + (b \times b)}$ (18) $\sqrt{(c \times c) - (b \times b)}$ (19) $\sqrt{(c \times c) - (a \times a)}$ (20) $\sqrt{\dfrac{V}{.7854 \times h}}$

Self-Check Test 26, pages 70 and 71
(1) 4 x 8 (2) (2 x 12) + (2 x 7) (3) 3.14 x 10 (4) 2 x 3.14 x 6.5 (5) 1.414 x 12.75 (6) 8.875 x 8.875 (7) 14.5 x 6.25 (8) $\frac{1}{2}$ x 24 x 14 (9) .433 x 11 x 11

(10) 2.598 x 15 x 15 (11) 2 x 3.14 x 8 x 5 (12) 4 x 3.14 x 7 x 7 (13) $\frac{1}{2}$ x 10 x (12 + 16) (14) 3.14 x 9 x 9 x 5.5 (15) $\sqrt{125}$ (16) $\sqrt{\dfrac{628.32}{.7854}}$
(17) $\sqrt{(10 \times 10) + (6 \times 6)}$ (18) $\sqrt{\dfrac{431.97}{.7854 \times 10}}$ (19) $\dfrac{2 \times 288}{18 + 24}$ (20) $\dfrac{2 \times 75}{18.75}$

Self-Check Test 26, pages 73 and 74
(1) 196 sq. in. (2) 254.469 sq. in. (3) 157.08 cu. in. (4) 22.5 in. (5) 6.51 in.

Self-Check Test 27, page 80
(1) 808 sq. ft. (2) 40,804 sq. ft. (3) 285.628 sq. ft. or 285.6 $7\frac{1}{2}$ in. (4) 1,531.25 sq. ft. or $1.531\frac{1}{4}$ sq. ft. (5) 2,332 tiles (6) approximately 20,000 tiles

Self-Check Test 28, page 82
(1) 448 ft. (2) 12,288 sq. ft. (3) 160 ft. (4) 2,700 tiles (5) 1104 sq. ft. (6) 168 sq. ft. (7) 936 sq. ft.

Self-Check Test 29, pages 83 and 84
(1) 96 sq. in. (2) 420 sq. in. (3) 132.3125 sq. in. or 132.3 sq. in. (4) 15.125 sq. ft. (5) 21.796875 sq. in. or 21.8 sq. in. (6) 60 sq. in. (7) 30 sq. in. (8) 44 sq. in. (9) 132 sq. in. (10) 320 sq. ft.

Self-Check Test 30, pages 85 and 86
(1) 49 sq. ft. (2) 595 sq. in. (3) 280 sq. in. (4) 28,600 sq. ft. (5) 420 sq. ft.

Self-Check Test 31, page 88
(1) 93.5 sq. in. (2) 5 ft. (3) 16.2375 sq. ft. or $16\frac{1}{4}$ sq. ft. (4) 15 ft. (5) 52,655 tiles

Self-Check Test 32, page 91
(1) 126 sq. ft. (2) 9,742.5 sq. ft. or $9,742\frac{1}{2}$ sq. ft. (3) $416\frac{2}{3}$ sq. ft. (4) approximately 2,660 tiles (5) 96 sq. ft.

Self-Check Test 33, page 94
(1) 11 ft. 4 in. (2) 7 ft. 4 in. (3) $12\frac{1}{2}$ sq. ft. (4) 428.31 sq. ft. (5) 12.5 sq. ft. (6) 28.6 sq. ft. (7) 481.9 sq. ft.

Self-Check Test 34, pages 99 and 100
(1) 47.124 in. or $47\frac{1}{8}$ in. (2) 380 sq. in. (3) 10.18 in. or $10\frac{3}{16}$ in. (4) 14.09 in. or $14\frac{3}{32}$ in. (5) 12.57 sq. ft. or 12 sq. ft. 82 sq. in. (6) 380.13 sq. ft. (7) 268.8 sq. ft. or 269 sq. ft. (8) 8,588.35 sq. in. (9) 371.54 sq. in. or $371\frac{1}{2}$ sq. in.

Self-Check Test 35, page 102
(1) 14 ft. $5\frac{7}{32}$ in. (2) 32.8 sq. or 33 sq. ft. (3) 377 sq. ft. (4) $35\frac{1}{8}$ sq. ft.

Self-Check Test 45, pages 131 and 132

(1) 1,331 cu. in. (2) 166.375 cu. ft. or $166\frac{3}{8}$ cu. ft. (3) 7,500 gal.
(4) 14.2 lb. (5) 22.9 cu. ft. (6) 205,000 cu. ft. (7) 320 cu. yd. (8) 106,875 gal. (9) 3,792.5 cu. yd.

Self-Check Test 46, page 134

(1) 18.85 cu. yd. or 19 cu. yd. (2) about 10,603 gal. (3) 43.63 cu. yd. or 44 cu. yd. (4) about 27 cu. ft.

Self-Check Test 47, page 136

(1) 1,650 cu. in. (2) 4,050 cu. ft. (3) $298\frac{2}{3}$ cu. yd. (4) 8789 gal.

Self-Check Test 48, page 138

(1) 60.6 cu. yd. (2) about 4639 gal. (3) about $\frac{1}{4}$ cu. yd.

Self-Check Test 49, page 139

(1) approximately 48 cu. ft. (2) approximately 493 cu. ft. (3) 146.7 cu. ft. or 147 cu. ft.

Self-Check Test 49, page 142

(1) 4,787.8 cu. in. or 4,788 cu. in. (2) 1,143.5 cu. ft. (3) approximately 1,094 cu. ft.

Self-Check Test 50, page 144

(1) 323.6 cu. in. (2) 44,731 gal. (3) 1,326.8 cu. in. (4) 18,938 cu. yd.

Self-Check Test 51, pages 148 and 149

(1) 10 bd. ft. (2) 24 bd. ft. (3) 75 bd. ft. (4) 24 bd. ft. (5) 360 bd. ft. (6) 1,792 bd. ft. (7) 5,000 bd. ft. (8) $298\frac{2}{3}$ bd. ft. or 299 bd. ft.

Self-Check Test 52, page 152

(1) 1,008 bd. ft. (2) 400 bd. ft. (3) 90 bd. ft. (4) 101.25 bd. ft. or $101\frac{1}{4}$ bd. ft.

Self-Check Test 53, page 154

(1) 56 sq. ft. (2) 336 sq. ft. (3) 96 sq. ft.

Self-Check Test 54, page 156

(1) 126 ft. Order six 22-foot lengths (2) Order eleven 16-foot lengths

Self-Check Test 55, page 157

(1) 3 pieces (2) 5 pieces (3) 6 pieces (4) 5 pieces (5) 4 pieces (6) 4 pieces

ANSWER KEY FOR RELATED MATHEMATICS FOR CARPENTERS—Continued

Self-Check Test 36, page 104

(1) 125.7 sq. in. (2) 1005.3 sq. ft. (3) 412.3 sq. in.

Self-Check Test 37, page 106

(1) 4 ft. $11\frac{11}{32}$ in. (2) 4.45 in. or $4\frac{7}{16}$ in. (3) 5 ft. $3\frac{7}{8}$ in.

Self-Check Test 38, page 108

(1) about 29 ft. $5\frac{7}{16}$ in. (2) 16.3625 sq. ft. or 16 sq. ft. 52 sq. in.
(3) approximately 49 sq. ft. (4) 5,236 sq. ft.

Self-Check Test 39, pages 113 and 114

(1) 36 sq. ft. (2) 1256 sq. in. (3) 541.5 sq. ft. or $541\frac{1}{2}$ sq. ft. (4) $12\frac{4}{9}$ sq. ft.
(5) 19.7 sq. ft. (6) 27 rolls (7) approximately 235 sq. ft. (8) 13.7625 gal.
(9) 1,666.75 sq. ft. or $1,666\frac{3}{4}$ sq. ft.

Self-Check Test 40, page 116

(1) 62.3 sq. ft. (2) 1,319.47 sq. ft. or $1,319\frac{1}{2}$ sq. ft. (3) 1,201.66 sq. ft. or 1,202 sq. ft. (4) 322.9 sq. ft. or 323 sq. ft.

Self-Check Test 41, page 118

(1) 450 sq. ft. (2) 117 sq. ft. (3) 324 sq. ft. (4) 605.8 sq. ft. or 606 sq. ft.

Self-Check Test 42, page 120

(1) 628.32 sq. ft. or $628\frac{1}{3}$ sq. ft. (2) 353.43 sq. ft. or $353\frac{1}{2}$ sq. ft.
(3) 94.248 sq. yd. or $94\frac{1}{4}$ sq. yd.

Self-Check Test 43, page 122

(1) 117 sq. ft. (2) 296 sq. ft. (3) 2,719.2 sq. in.

Self-Check Test 43, page 124

(1) 1,244.6 sq. in. or 1345 sq. in. (2) 1,791.89 sq. in. or 1,792 sq. in.
(3) 808.5 sq. in.

Self-Check Test 44, page 126

(1) 452.39 sq. in. (2) 1,256.64 sq. ft. or 1,257 sq. ft. (3) 245.4375 sq. in. or $245\frac{7}{16}$ sq. in. (4) 3,927 sq. ft.

ANSWER KEY FOR RELATED MATHEMATICS FOR CARPENTERS—Continued

Self-Check Test 55, page 158

(1) 300 bd. ft. (2) 3,584 bd. ft. (3) 150 bd. ft. (4) 1,600 bd. ft. (5) 432 bd. ft. (6) 512 bd. ft. (7) Total 2,544 bd. ft.

Self-Check Test 56, page 160

(1) 31 (2) 29 (3) 28

Self-Check Test 57, pages 167 and 168

(1) 32 pieces (2) 112 pieces (3) 11,662 in. (4) 12,806 in. (5) 17,205 in. (6) 18,439 in. (7) about 46 ft. (8) 172 ft. (9) 125 ft. (10) 226 sq. ft.

Self-Check Test 58, page 170

(1) 501.6 or 502 sq. ft. (2) 1112.8 or 1113 sq. ft. (3) 459.8 or 460 sq. ft. (4) 941.6 or 942 sq. ft.

Self-Check Test 59, page 174

(1) (a) 154 studs (b) 232 studs (2) 216 studs

Self-Check Test 60, page 178

(1) 396 ft. (2) (a) 1,320 feet (b) 440 feet of plates; 220 feet of ribbons

Self-Check Test 61, page 180

(1) 360 sq. ft. (2) 900 sq. ft. (3) 2,120 sq. ft.

Self-Check Test 62, page 188

(1) 693 sq. ft. (2) 672 sq. ft. (3) 3,240 sq. ft. (4) 1419 sq. ft. (5) 640 sq. ft.

Self-Check Test 63, page 190

(1) 24 ft. (2) 12 ft. (3) $\frac{1}{3}$ (4) 12 ft.

Self-Check Test 64, pages 196 to 198

(1) 21 ft. $9\frac{5}{8}$ in. (2) 21 ft. (3) 22 ft. $2\frac{1}{2}$ in. (4) 29 ft. $6\frac{7}{8}$ in. (5) 19 ft. $11\frac{11}{16}$ in. (6) 15 ft. $9\frac{3}{4}$ in. (7) 27 ft. $10\frac{7}{8}$ in. (8) 43 rafters (9) 180 rafters (10) 80 rafters

Self-Check Test 65, page 202

(1) 531 sq. ft. (2) 199 sq. ft. (3) approximately 682 sq. ft. (4) about $4\frac{1}{2}$ squares or 18 bundles (5) about $6\frac{1}{4}$ squares or 25 bundles

Self-Check Test 66, page 204

(1) Approximately 10 square (2) 40 bundles (3) 19 bundles (4) Approximately $3\frac{1}{2}$ squares (5) Approximately 21 bundles

Self-Check Test 67, page 206

(1) 495 sq. ft. (2) 370 sq. ft. (3) 431 sq. ft. (4) 389 sq. ft. (5) 88 sq. ft.

Self-Check Test 68, pages 209 and 210

(1) 24 risers (2) 16 risers (3) 16 risers (4) 10 in. (5) $10\frac{1}{2}$ in. (6) 11 in. (7) $7\frac{5}{8}$ in. (8) 15 risers (9) $10\frac{1}{2}$ ft. (10) $8\frac{1}{8}$ in.